SOL-GEL METHOD - DESIGN AND SYNTHESIS OF NEW MATERIALS WITH INTERESTING PHYSICAL, CHEMICAL AND BIOLOGICAL PROPERTIES

Edited by **Guadalupe Valverde Aguilar**

Sol-Gel Method - Design and Synthesis of New Materials with Interesting Physical, Chemical and Biological Properties
http://dx.doi.org/10.5772/intechopen.76535
Edited by Guadalupe Valverde Aguilar

Contributors

Nelson Daniel Vejar, Maritza Paez, Roberto Solis, Evelyn Gonzalez, Maria Victoria Enciana, Lisa Muñoz, Habibollah Aminirastabi, guoli ji, Hao Xue, Dongliang Peng, Rabaah Syahidah Azis, Muhammad Syazwan Mustaffa, Sakinah Sulaiman, Emmanuel Ifeanyi Ugwu, Farida Selim, David Winarski, Guadalupe Valverde Aguilar

First published in London, United Kingdom, 2019 by IntechOpen
IntechOpen is the global imprint of INTECHOPEN LIMITED, registered in England and Wales, registration number: 11086078, The Shard, 25th floor, 32 London Bridge Street
London, SE19SG – United Kingdom
Printed in Croatia

British Library Cataloguing-in-Publication Data
A catalogue record for this book is available from the British Library

Additional hard copies can be obtained from orders@intechopen.com

Sol-Gel Method - Design and Synthesis of New Materials with Interesting Physical, Chemical and Biological Properties,
Edited by Guadalupe Valverde Aguilar
p. cm.
Print ISBN 978-1-78985-333-9
Online ISBN 978-1-78985-334-6

For EU product safety concerns:
IN TECH d.o.o., Prolaz Marije Krucifikse Kozulić 3, 51000 Rijeka, Croatia, info@intechopen.com or visit our website at intechopen.com.

Meet the editor

Professor Guadalupe Valverde Aguilar obtained her PhD. in Physics in 2003 at the Institute of Physics, UNAM. She undertook a Postdoctoral stay at UCLA, USA in 2003-2005. As a staff member of the UNAM, her research focused on spectroscopy techniques such as optical absorption, luminescence, photoluminescence, Raman spectroscopy and photoconductivity.
Professor Valverde has been a professor in Physics at the Instituto Politécnico Nacional since 2011 and is an experimentalist carrying out research on bulk materials and mesostructured sol-gel thin films produced by a combination of the sol-gel process and evaporation-induced self-assembly (EISA). Her research at the CICATA U. Legaria is focused on catalysis and photocatalysis, release of drugs for the treatment of Parkinson's disease, and magnetism. This research is carried out in a highly interactive environment and involves inter-collaboration at ESIQIE-IPN, UNAM, and international collaboration with colleagues in USA.

Contents

Preface

In 1998, I learned the sol-gel method to produce rhodamine 6G monoliths. For this, as any beginner, I consulted the work of Professor Jeffrey Brinker, *Sol-gel science: the physics and chemistry of sol-gel processing*, to learn and understand the principles of this wonderful method of synthesis, which I continue to use in my current research. This was the beginning of an exciting journey in the field of sol-gel, which allowed me to design and produce novel materials for various applications in magnetism, non-linear optics, photoconductivity and photoluminescence.

Subsequently, I had the opportunity to work with Prof. Jeffrey I. Zink (Faculty Distinguished Professor of the Department of Chemistry, University of California, Los Angeles, USA), an expert and pioneer in the sol-gel synthesis of mesoporous silica supports with lots of research publications on the subject, who gave me his knowledge about this field.

Currently the sol-gel method is one of the most used synthesis methods as it allows for the preparation of an infinite number of materials and ceramics. Depending on the application, it allows us to design our materials at different scales, micro-, meso- and macro-; and in different forms such as film, fiber, glass, and powder. The versatility of this method allows us to explore different areas of knowledge that cover global problems such as energy, biotechnology, and electronics. Sol-gel materials can be of different kinds: catalysts, nanocarriers, magnetic and metallic nanoparticles, etc. As already mentioned, the advantages of the sol-gel method are its versatility for the design of different materials at low temperatures, and low production cost.

Other synthetic routes have emerged, but this has not diminished the popularity of the sol-gel method and it remains one of the most used methods in the design and control of different materials due to its great flexibility to control the properties based on the application that you want. As a consequence of this, sol-gels have a wide variety of applications that are reported in fields such as photocatalysis, biomedicine, drug release, treatment of neurodegenerative diseases, magnetism, optoelectronics, etc.

We hope that the material contained in this book is of interest to the reader, and we look forward to your enthusiastic participation in future projects. I greatly appreciate the authors for their invaluable time and interest in the preparation of each of their contributions.

Prof. Guadalupe Valverde Aguilar
Instituto Politécnico Nacional
CICATA Unidad Legaria
Mexico City, Mexico

Chapter 1

Introductory Chapter: A Brief Semblance of the Sol-Gel Method in Research

Guadalupe Valverde Aguilar

Additional information is available at the end of the chapter

http://dx.doi.org/10.5772/intechopen.82487

1. Introduction

The emergence of the sol-gel process occurred in the year 1921. In the 1960s, its development was given due to the need of new synthesis methods in the nuclear industry. This development began to become popular around 1984 and reached its splendor in 2011 as shown in **Figure 1** [1]. Dr. Jeffrey Brinker is a pioneer in the synthesis of materials and sol-gel-science and sets an

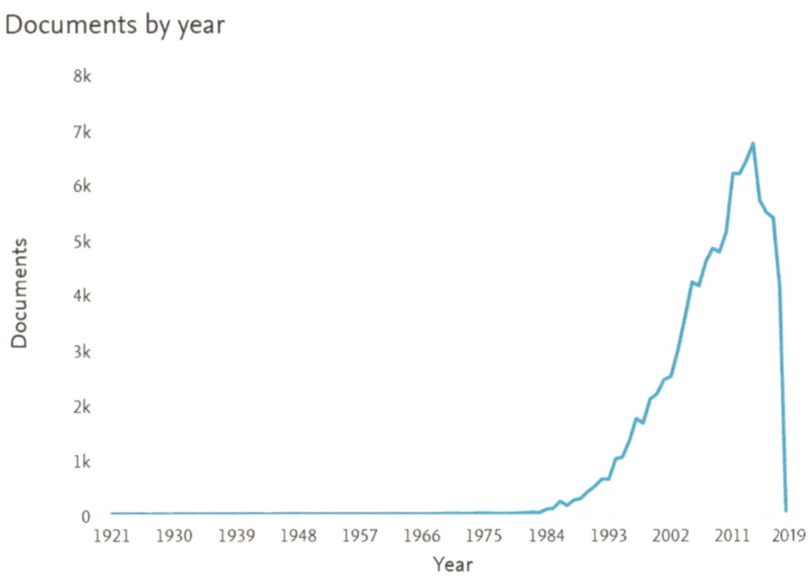

Figure 1. Temporal evolution of sol-gel publications.

important guideline for the flowering of the sol-gel method [2]. Other researchers who have developed their research in the sol-gel field making substantial and important contributions to this field are Dongyuan Zhao and David Avnir. The method of sol-gel is a route of synthesis more used worldwide that there is a meeting named International Sol-Gel Conference [3], which is held every 2 years, bringing together renowned scientists with the new generations, who give contributions of their work in different areas, all linked to the sol-gel method.

2. General mechanism

A description of the sol-gel process can be formation of an oxide network through polycondensation reactions of a molecular precursor in a liquid.

In general, in this process, several stages are identified, starting with a silicate solution and then forming a sol, which will then be transformed into a gel, and finally, a dry gel is obtained which is generally formed by a three-dimensional network of silica, with numerous pores of various sizes interconnected. **Figure 2** presents an outline of the routes of this mechanism.

Among the advantages of using the sol-gel process in the synthesis is because it can be carried out at room temperature, it allows us to produce a wide range of novel and functional materials, with potential applications in different areas; and finally, it is really attractive compared to other methods, due to its low production costs.

Sol-gel samples can be designed with a wide variety of morphologies, such as monoliths, films, fibers, and powders. In particular, films are the most important from the technological point of view.

The process begins with the formation of a "sol," which is a stable dispersion of colloidal particles (amorphous or crystalline) or polymers in a solvent. A "gel" is formed by a three-dimensional continuous network, which contains a liquid phase, or by the joining of polymer chains. In a colloidal gel, the network is built from agglomerates of colloidal particles. While in a polymer gel, the particles have a polymeric substructure composed of aggregates of sub-colloidal particles. Generally, van der Waals forces or hydrogen bonds dominate the interactions between the sol's particles. During synthesis, in most gel systems, covalent-type interactions dominate, and the gel process is irreversible. The gelation process may be reversible if there are other interactions involved [2].

The purpose behind the sol-gel synthesis is to dissolve a compound in a liquid to obtain a solid controlling the factors of said synthesis. Using a controlled stoichiometry, sols of different reagents can be mixed to prepare multicomponent compounds. The sol-gel method prevents the problems with coprecipitation, which may be inhomogeneous, as it is a gelation reaction. It allows mixing at an atomic level to form small particles, which are easily sinterable.

Typically, in the sol-gel chemistry, there is a reaction of an organometallic compound, which is generally an alkoxide, nitrate, or chloride under aqueous conditions to form a solid product. This product can be a dense glass monolith, a high surface area molecular filter, an aerogel to

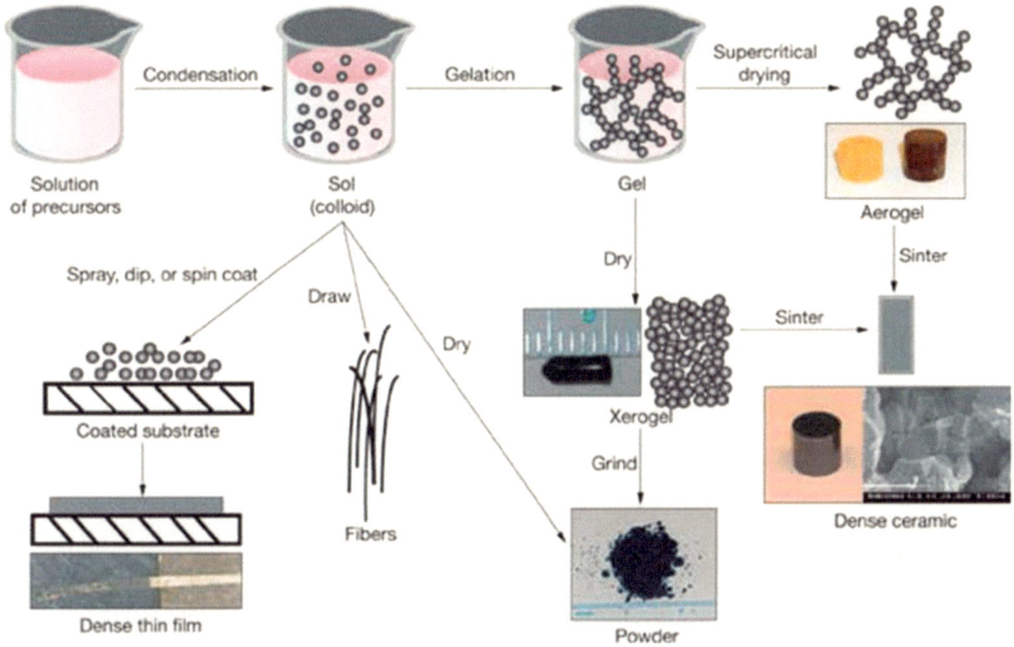

Figure 2. Stages of the sol-gel process.

a metal oxide, a nitride coating, or nanoparticle. The process begins with reactions of hydrolysis and condensation of a precursor to form a gel followed by aging, solvent extraction, and finally drying. These reactions may be catalyzed by the addition of an acid or a base, which will produce dense or diffuse networks, respectively, by altering the hydrolysis kinetics. The selection of the precursor and catalyst depends ultimately on what you would like to make [4].

In the gelation step, condensations are produced from the gel precursors in aqueous solution which are hydrolyzed and polymerized through alcohol or water. When starting the gelation, when the average size of the conglomerate is very small, they are best modeled with an approximation at the atomic level.

In the last decades, a remarkable effort has been made to develop theoretical models for this, with convincing results. In the case of hierarchically structured gels and low density gels, these cannot be analyzed directly with molecular models; a mesoscale approach should be used. In contrast, relatively dense gels can be modeled with simulations at the atomic level or coarse grain simulations. Aging of the gel is an extension of the gelation step in which the gel network is reinforced by an additional polymerization, which can be controlled by varying the temperature and the type of solvent.

In the next stage, syneresis can occur during the aging of the gel, which is the expulsion of solvent due to the contraction of the gel matrix. The process of drying the gel consists in eliminating the water from the gel system, with simultaneous collapse of the gel structure, under conditions of constant temperature, pressure, and humidity [5, 6].

Usually, the dry gel is given a calcination treatment to turn it into a crystalline material. The following reactions usually occur: desorption of solvent and water physically absorbed from the walls of micropores (100–200°C), decomposition of residual organic groups into carbon dioxide (300–500°C), collapse of small pores (400–500°C), collapse of larger pores (700–900°C), and continued polycondensation (100–700°C). The phenomena of sintering and densification are produced through different mechanisms such as condensation by evaporation, surface diffusion, grain limit, and mass diffusion.

3. Design of sol-gel materials

As mentioned earlier, the sol-gel method allows the preparation of an infinity of materials and ceramics. Its great versatility allows us to cover different areas of knowledge that cover global problems such as energy, biotechnology, electronics, health, pollution, scaffolds for tissue engineering, and smart coatings molecular imprinting [7–10]. In this way, sol-gel materials can be of different kinds, since catalysts, nanocarriers, inorganic pigments, drugs, magnetic and metallic nanoparticles. Also, it allows the encapsulation of biological molecules such as proteins and enzymes [11–13] which have applications as biosensors or the release of drugs in the treatment of neurodegenerative diseases such as cancer, Parkinson's, and Alzheimer's, for example.

Sol-gel chemistry tends to be particularly sensitive to the following parameters:

- pH: any colloidal chemistry that involves water is sensitive to pH.
- Solvent: in the polymerization process, as molecules are assembled into nanoparticles, the solvent plays two important roles; the first is that it must be able to keep the dissolved nanoparticles so that they do not precipitate out of the liquid; and second, it must play a role in helping nanoparticles connect with each other.
- Temperature: the chemical kinetics of the different reactions involved in the formation of nanoparticles and the assembly of the nanoparticles in a gel network are accelerated with temperature, which affects the gel time. At very low temperatures, gelation is a slow process that can take weeks or months. In contrast, at high temperatures, the reactions that bind the nanoparticles to the gel network occur so quickly that lumps form in their place and a solid precipitates out of the liquid. The gelation temperature must be controlled to optimize the reaction time.
- Time: depending on the type of gel to be obtained, the different steps in the gel formation process work differently at different time scales. In general, it is recommended that the formation of the gel should be slow to produce a very uniform structure, resulting in a stronger gel. Accelerating reactions through short times cause precipitates to form instead of gel network and can cause a gel to become cloudy and weak or simply not form.
- Catalysts: a chemical reaction can be accelerated by the presence of a catalyst. In much of the sol-gel chemistry, this is very pH sensitive. This is because both acids (H^+) and bases (OH^-) are catalysts but accelerate chemical reactions by different mechanisms.

- Agitation: at this stage, the mixing of the sol during gelation should ensure that the chemical reactions in the solution are produced uniformly, allowing all molecules to receive an adequate supply of the chemicals they need for these reactions to be carried out correctly. Generally, there are microscopic and macroscopic domains of gel networks partially formed throughout the liquid, and agitation can sometimes break up the formation of these domains; and the network fragments grow back into a wider network.

Therefore, taking into account these factors and the type of application, many protocols have been used to design our materials in different scales, nano-, micro-, meso-, and macromaterials, all aimed at optimizing and maximizing their optical, electrical, magnetic, and nonlinear properties [7, 8, 14]. It is described how these factors influence said properties during sol-gel reaction [7].

In this work, valuable contributions in different fields related to novel materials synthesized by the sol-gel route are shown, all with topics of great technological importance and which have an impact on engineering applications, at the level of electronics, health, and coatings.

Author details

Guadalupe Valverde Aguilar

Address all correspondence to: mvalverde@ipn.mx

Department of Nanotechonology and Functional Materials, CICATA Unidad Legaria, Instituto Politécnico Nacional, Mexico

References

[1] https://www.scopus.com/search/form.uri?display=basic

[2] Brinker CJ, Scherer G. Sol-Gel Science. New York: Academic Press; 1989. ISBN 9780080571034

[3] http://solgel2019.ifmo.ru/

[4] Collins A. Nanotechnology Cookbook: Practical, Reliable and Jargon-Free Experimental Procedures. UK, Elsevier Science; 2012. ISBN-13: 978-0080971728

[5] Hench LL, West JK. The sol-gel process. Chemical Reviews. 1990;**90**:33-72

[6] Sakka S, Kozuka H. Handbook of Sol-Gel Science and Technology. 1. Sol-Gel Processing. The Netherlands: Kluwer Academic Publishers; 2004. ISBN 1-4020-7966-4

[7] Lee BS, Lin HP, Chan JCC, Wang WC, Tsai YH, Lee YL. A novel sol-gel derived calcium silicate cement with short setting time for application in endodontic repair of perforations. International Journal of Nanomedicine. 2018;**13**:261-271

[8] Mokhtari K, Salem SH. A novel method for the clean synthesis of nanosized cobalt based blue pigments. RSC Advances;**217**(7):29899

[9] Shi C, Ding GS, Tang AN, Qiao YY. Synthesis and evaluation of ion-imprinted sol-gel material of selenite. Analytical Methods. 2017;**9**:1658-1664

[10] MRD K, Shafeeyan MS, AAA R, WMAW D. Application of doped photocatalysts for organic pollutant degradation—A review. Journal of Environmental Management. 2017;**198**:78-94

[11] Karataş A, Algan AH. Template synthesis of tubular nanostructures for loading biologically active molecules. Current Topics in Medicinal Chemistry. 2017;**17**:1555-1563

[12] Araújo-Gomes N, Romero-Gavilán F, Sánchez-Pérez AM, Gurruchaga M, Azkargorta M, Elortza F, et al. Characterization of serum proteins attached to distinct sol–gel hybrid surfaces. Journal of Biomedical Materials Research—Part B Applied Biomaterials. 2018; **106**:1477-1485

[13] Gill JK, Orsat V, Kermasha S. Screening trials for the encapsulation of laccase enzymatic extract in silica sol-gel. Journal of Sol-Gel Science and Technology. 2018;**85**:657-663

[14] Ben-Arfa BAE, Miranda Salvado IM, Ferreira JMF, Pullar RC. Novel route for rapid sol-gel synthesis of hydroxyapatite, avoiding ageing and using fast drying with a 50-fold to 200-fold reduction in process time. Materials Science and Engineering C. 2017;**70**:796-804

Chapter 2

The Effect of Annealing, Doping on the Properties and Functionality of Zinc Oxide Thin Film; Review

Emmanuel Ifeanyi Ugwu

Additional information is available at the end of the chapter

http://dx.doi.org/10.5772/intechopen.79018

Abstract

The review of the effect of annealing and doping zinc oxide thin films on both the structural and optical properties has been carried out for different growth techniques such as sol–gel growth technique. The structural and optical properties were carried out using thin films were characterized SEM, XRD while TE and TM guided mode spectra, UV–VIS–NIR (HR4000Ocean Optics) and UV–Visible spectrometry were used accordingly respectively. From the results, it was clearly observed the both the morphological and the crystal characteristics structural characteristic, although increase in the percent of doping element affected it as the diffraction peak was shifts slightly to a lower angle side with report that crystal structure of the film deteriorate at a higher doping concentration of doping element as it decreases the c-lattice. There was also adjustment on the band gap of the material when it was annealed at various temperatures and also when the doping concentration was varied. The film exhibited lower absorbance, high transmittance depend on the regions of electromagnetic wave spectra.

Keywords: zinc oxide film, sol–gel deposition, chemical bath deposition doping, annealing, structural and optical properties, morphology, characteristics, analysis, absorbance, transmittance, band gap

1. Introduction

The aim of this review work is to overview the rapid progress of thin film techniques to grow ZnO based thin film which has been on course for long and to view how doping and annealing using different growth techniques affects its characteristics and functionality.

Zinc Oxide, ZnO thin film is one of the most oxide based thin film materials of the II-VI semiconductors are being studied since early twentieth century with great interest by non-scientists world-wide [1]. Now due to its current applicability to several novel devices, electronics, opto-electronics there has been a renewed attention are being given to this material thin film [2, 3].

At various point in time, conferences had been held and the proceedings published exclusively for ZnO thin film in some places such as Singapore and Changchan in China during 2005, 2009 and 2005 respectively based on exploring the efficacy and potentiality on the applicability of the thin film in order to create awareness on the feasibility of commercial application of the thin film for feature devices. Yet it seems as if the realm of the novel devices from this wonderful and unique oxide based thin film material is yet to be actualized in full [4]. With a direct band gap of 3.2 eV and large excitation energy of 60 meV at room temperature ZnO thin film just like GaN is a good candidate for blue and ultraviolet-optical devices [5]. Though it has more excitation energy and wider band gap than GnN, it can be grown into single crystal on the substrates with its good broad chemistry which leads to opportunities for wet chemical etching, low power threshold for optical pumping, radiation hardness and biocompatibility, thermal stabilities and environmental friendly nature [6]. Crystallographically, ZnO thin film has a hexagonal closed packed structure, wurtizite type with the zinc and oxygen ionic plane stacked alternatively along the principal axis of the symmetry which made to it have an excellent piezoelectric and optical properties [7–9]. The flexibility of the its heterostructures has been found to lead to expanded possibility of its device functionalities to various kind of application apart from solar energy devices. Uniquely apart from its easy realization of bipolar based devices due to doping symmetry issue which characterizes other II-IV semiconductors that can be readily doped n-type, it has been found very difficult to reproduce the ZnO doped p-type semiconductor because of lack of dopants having shallow acceptor level as a result of low dopant stability [10]. The dopants also affects in a strong term the microstructural, electrical and optical characteristics of the thin film [11, 12]. Glaringly, it has been ascertained the doping an annealing optimizes the solid state and optical characteristics of ZnO thin film [13–18]. It has been reported that doping do not only affect the optical properties of ZnO thin film but also the physical properties [11, 19].

In this paper we wish to review how doping and annealing of zinc oxide thin film affects the structural and optical characteristics and how it restructures the properties for utilization in optoelectronic and solar energy application.

2. Materials and methods

Various methods has been used to grow ZnO thin film at different time and places such as chemical bath deposition, sol–gel growth techniques, radio-frequency magnetron sputtering and chemical vapor deposition under low vacuum condition, etc. In the case of CBD, different complex agents such as ammonia, hydrazine, ethanolamine, methylamine, triethanolamine, etc. has been used to deposit the thin film both at room temperature and at annealed temperature ranging from 150 to 400°C [19–21] while spray pyrolysis using aqueous methanolic solution zinc acetate as spraying solution has been used to develop the thin film apart from the use of ultrasonic spray pyrolysis technique that was carried out at 200 to 500°C by many researchers [3, 22]. Apart from atomic layer deposition technique that has been reported a good candidate for the growth of high performance n-type growth of ZnO at low temperature sol–gel technique has not been has been employed using $Zn(CH_3COO)_2.2H_2O$ as a starting material to prepare an acetone gas sensor for use for the growth of the thin film temperature ranging between 700 and 800°C. ZnO thin film has been doped with different elements using different growth techniques such as Sol–gel, Magnetron sputtering, CBD, Spray CVD etc. [23–27]. The morphological structures for both as-deposited, annealed and doped ZnO based thin films were analyzed using SEM of different models such as Hitachi-S4500, JEOL6390-LV respectively; and the same for the optical characterization, different models of spectrometers such as TE and TM guided mode spectra, UV–VIS–NIR (HR4000Ocean Optics) and UV–visible spectrometry were used accordingly respectively.

Other optical parameters such band gap, extinction coefficient and optical conductivity were computed using the relations [28, 29].

$$(h\alpha\upsilon)^2 = C\left[h\upsilon - E_g\right] \tag{1}$$

$$k = \frac{\alpha\lambda}{4\pi} \tag{2}$$

$$\sigma_o = \alpha nc\,\varepsilon_o \tag{3}$$

where the α is absorption coefficient that is dependent on the photon energy and c and n are the velocity of light and the refractive index of the film respectively.

3. Results/discussion

3.1. Morphological analysis

The binding energy of the ZnO thin film deposited by single source chemical vapor deposition technique was analyzed with the region XPS spectra and was found to be about 532.2 eV which was attributed to the effect of zinc hydroxide on the surface of the film [30]. Similarly, the value obtained for the same thin film prepared by chemical spray pyrolysis on silicon using wide scan XPS on the deposited thin film depicted high value 530.19 and 531.82 eV respectively due to O—Zn bond and O—H bond absorbed water molecule [22]. Contrarily the theoretical excitation energy proposed for the same thin film in the literature was just 60 meV at room temperature. The morphological characterization was carried out by SEM. The observation as in **Figures 1** and **2** indicated that the surface and cross-section morphologies of pure ZnO thin films has smooth surface and dense polycrystalline microstructure in the form of small grain with increase in particle size at high annealing temperature while in the case of **Figure 3**, the Te-doped SEM micrographs it was found that the grains were oriented and larger than those observed in pure ZnO. However, in Cu-doped ZnO thin film the morphology depicted a decrease in grain size with percentage increase in copper concentration with annealing affected the microstructural properties while in case of aluminum doped ZnO thin film it was seen that the increase in Al concentration alone led to a significant increase in improvement of the crystal compactness of the structure as in **Figure 4**.

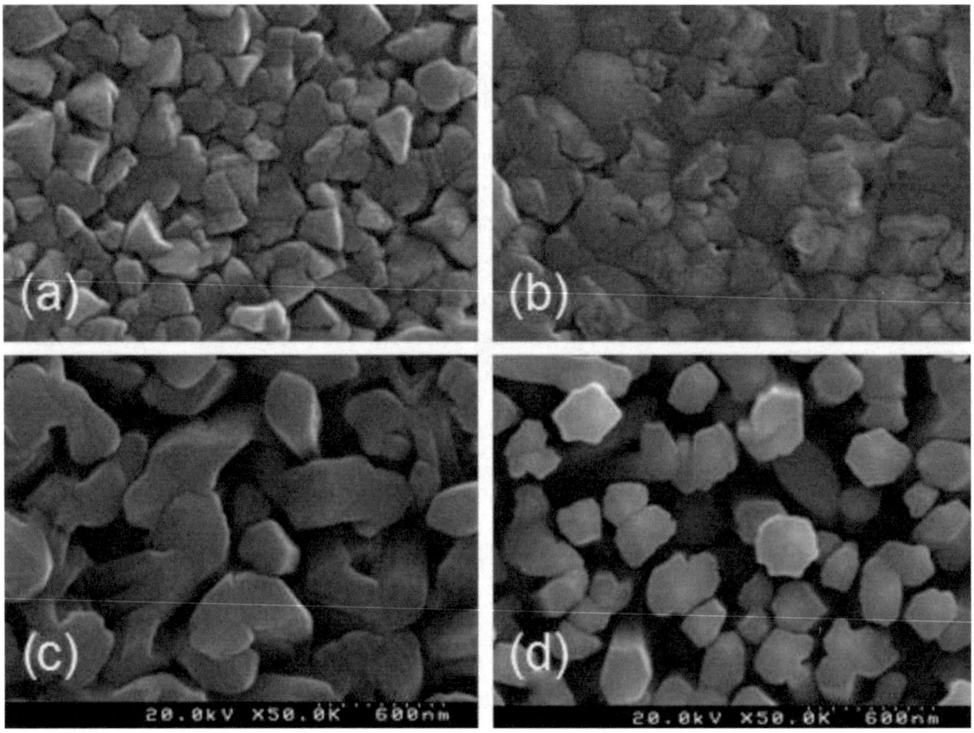

Figure 1. SEM surface morphology [a–d] of pure ZnO thin film annealed at temperatures of 100, 200, 300 and 400°C respectively.

3.2. Structural analysis

The XRD analysis allows us to determine the crystal orientation of both the as-deposited, doped and annealed ZnO thin film on glass substrates. The diffraction patterns of the samples for all these mentioned cases are depicted in **Figures 5–9** respectively and from the analysis, it was seen that as-deposited ZnO thin film had common intense peaks irrespective of time of growth and substrate temperatures occurring within $2\theta = 32°$ and $36°$ along the following orientations

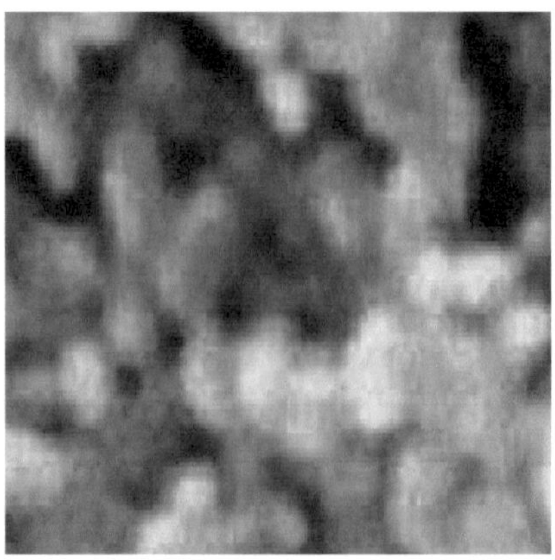

Figure 2. SEM morphological structure of undoped ZnO thin film at magnification ×50,000.

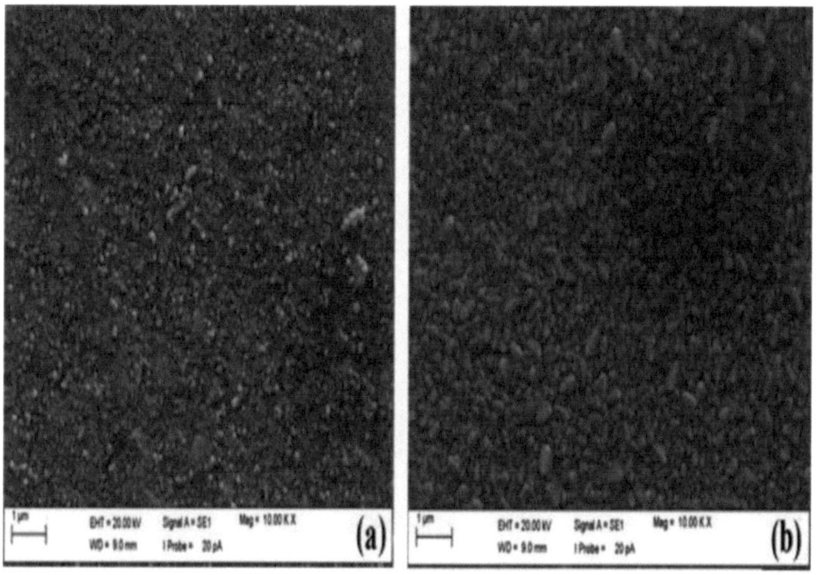

Figure 3. SEM morphological structures of as deposited and Tellurium doped ZnO thin film (a and b).

(100), (002) and (101) as in **Figures 5** and **6**. In the case of Cu-doped and boron doped ZnO thin film, intense peaks were observed at (100), (002),(101 and (102) respectively all occurring within 30° and 40° with increase in the intensity of the peak as annealing temperature increases as in

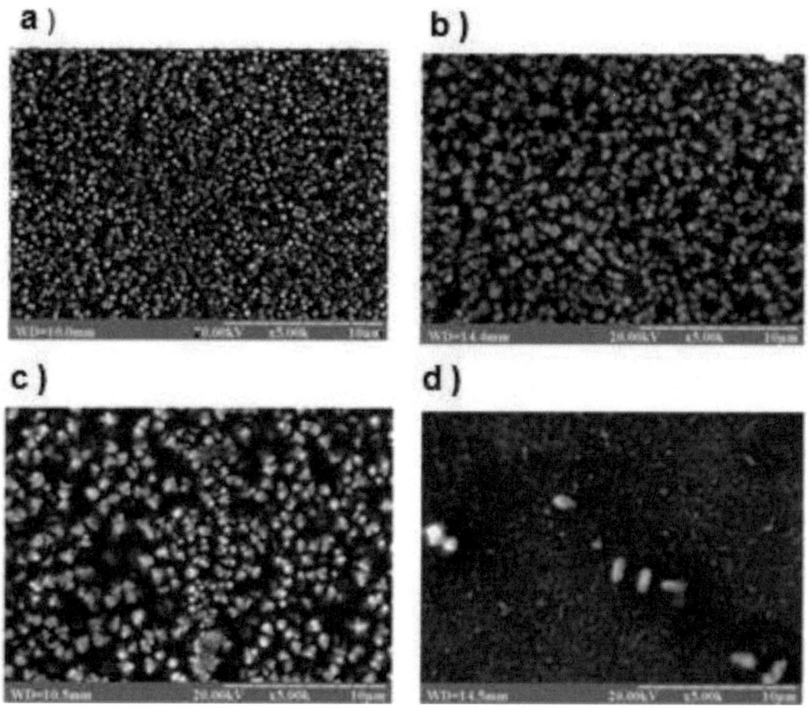

Figure 4. SEM morphological structure of Cu-doped ZnO thin film [a–d] as deposited and annealed at temperature; 500, 700 and 850°C.

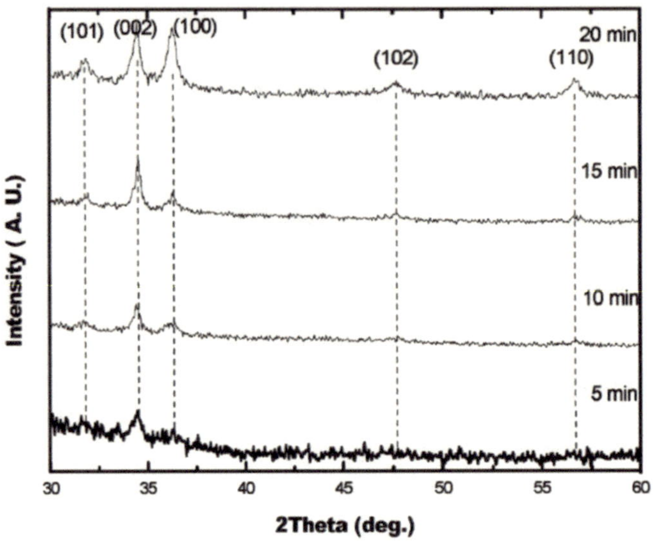

Figure 5. XRD spectra for pure and [1, 3, 5] % Al doped ZnO thin films.

Figure 7. The XRD for boron doped ZnO as in **Figure 7** exhibited defined intense peaks along (100),(002), (110) (103), (200) and (112) within 32° to 37° and 48° and 67° respectively while for Al-doped, intense peaks were identified along (100), (002) and (101) at 2θ = 36.24, 32.37 and 36.24° respectively. Generally it was seen that irrespective of annealing growth technique and doping not withstanding doping element used, ZnO has high diffraction peaks elaborated at (100) and (002) in all cases This observation indicated that ZnO thin film is strongly c-axis oriented with wurtizite structural characteristic, although increase in the percent of doping element affects and often shifts the diffraction peak slightly to a lower angle side with report that crystal structure of the film deteriorate at a higher doping concentration of doping element as it decreases the c-lattice [10, 13, 25].

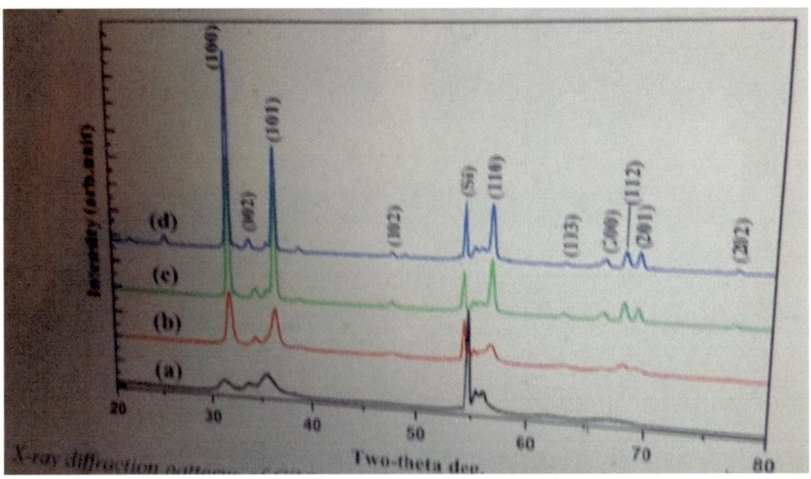

Figure 6. XRD spectra for pure ZnO thin film for different deposition time.

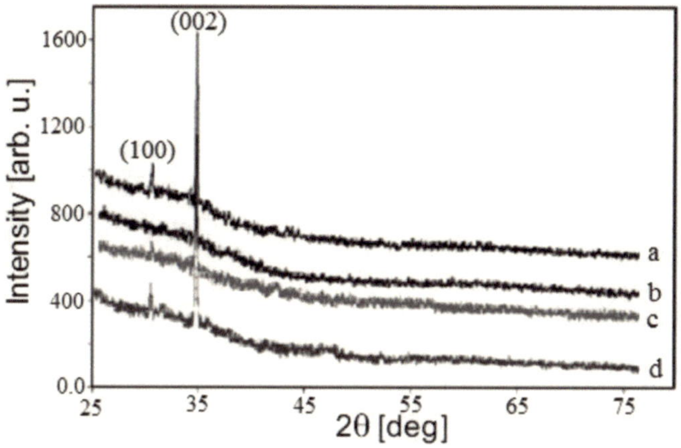

Figure 7. XRD spectra of pure and Cu-doped ZnO thin film annealed at temperatures; 500, 700 and 850°C.

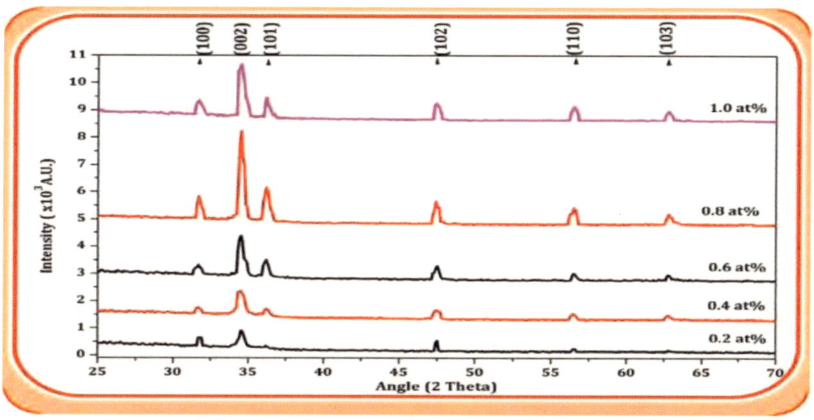

Figure 8. XRD spectra for various percentage boron doped ZnO thin film.

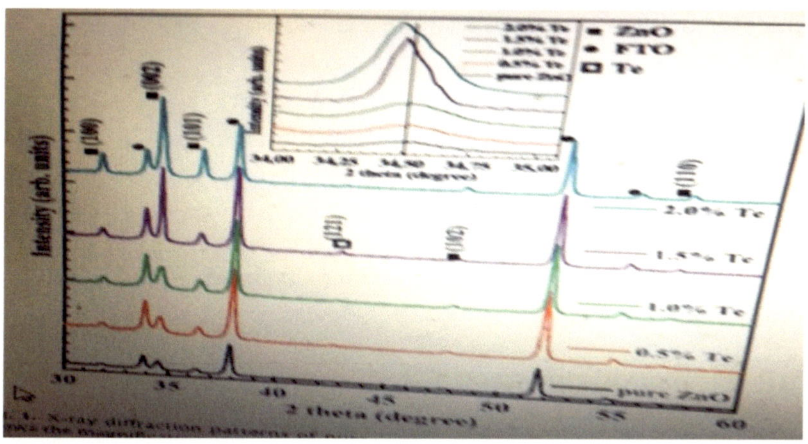

Figure 9. XRD spectra of pure and varying percentage tellurium doped ZnO thin film.

3.3. Analysis of the energy band gap

The band gap of ZnO thin film as recorded in the all the experiment for both as deposited, annealed and doped was based on Tauc model which involved a plot of a curve of $(\alpha h\upsilon)^2$ as function of photon energy, *hυ* (eV) [22]. In the plot, the band gap is obtained by extrapolating the straight line portion of the curve/tangential line to the photon energy axis from the extrapolation as in **Figure 10**, it was observed that the band gap for pure an annealed ZnO shifts/narrows from

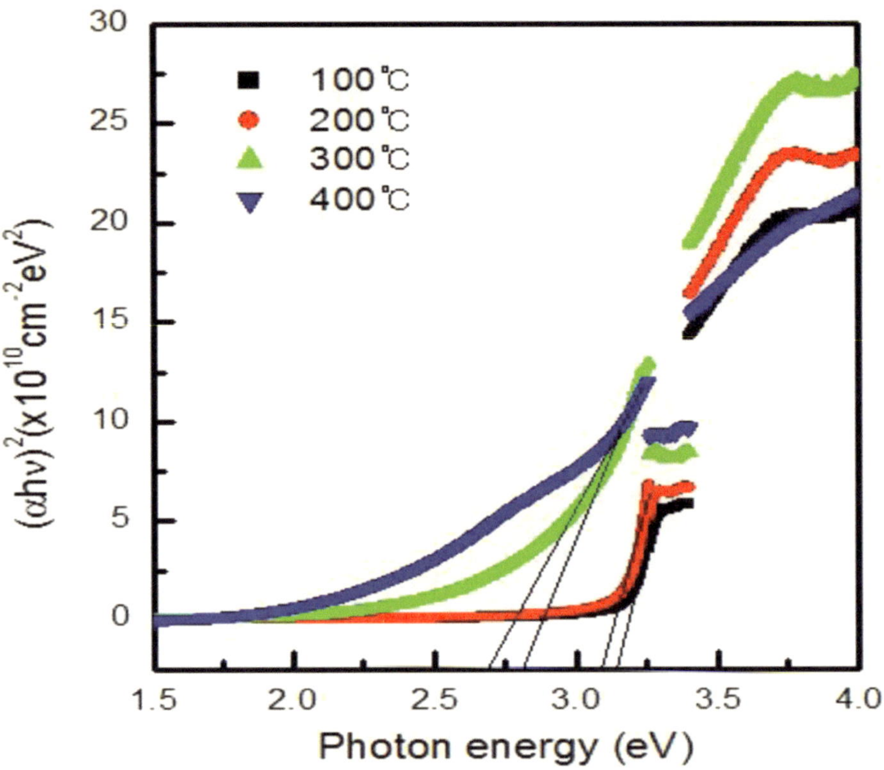

Figure 10. Graph of $(\alpha h\upsilon)^2$ as a function *hυ* for pure ZnO thin film grown under different temperature.

3.13 eV at 100°C to 3.09 eV at 200°C and finally to 2.69 eV at 400°C respectively. This is in accordance with Ayouchi report in his work [22]. Form the Tellurium doped ZnO thin film, using the same Tauc model, it was noted that the undoped film has its band gap as 3.18 eV the percentage doping concentration of Te increased the energy band gap up to maximum of 3.42 eV for 1.5% of tellurium doping concentration. This was as a result of transition energy degeneracy associated with semiconductor owing to the partially filled conduction band [30, 31, 32–34].

4. Absorbance/transmittance

The spectral absorbance of pure thin film deposited at different temperature as in **Figure 11** was found to be decreasing in slope slowly with a sharp sink observed within 779 nm after which there was decrease in the value of the absorbance the red region and infra-red region of electromagnetic wave spectrum.

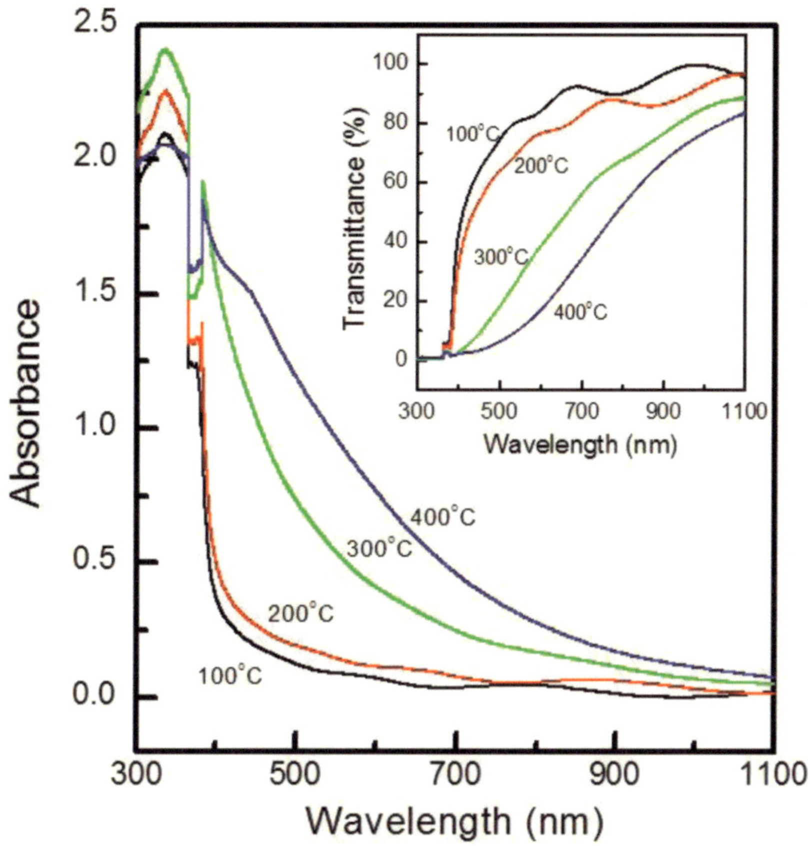

Figure 11. Spectral absorbance of pure ZnO thin film grown under different temperature.

The transmittance as observed in **Figures 11–13** for both pure annealed and doped ZnO thin films occurred within 600–1100 nm. From the report of ZnO doped with nitrogen had its transmittance increased between 395 and 590 nm and thereby remained constant within 80–90% around the infrared region [800–200 nm] which collaborated the results in the figures mentioned [27, 28].

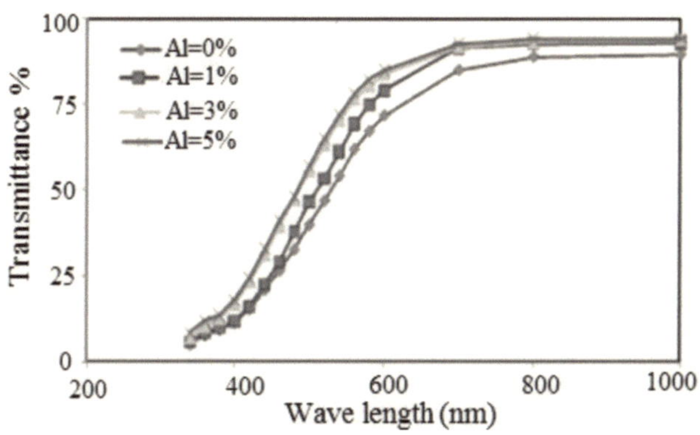

Figure 12. Percentage transmittance of ZnO dope with % aluminum.

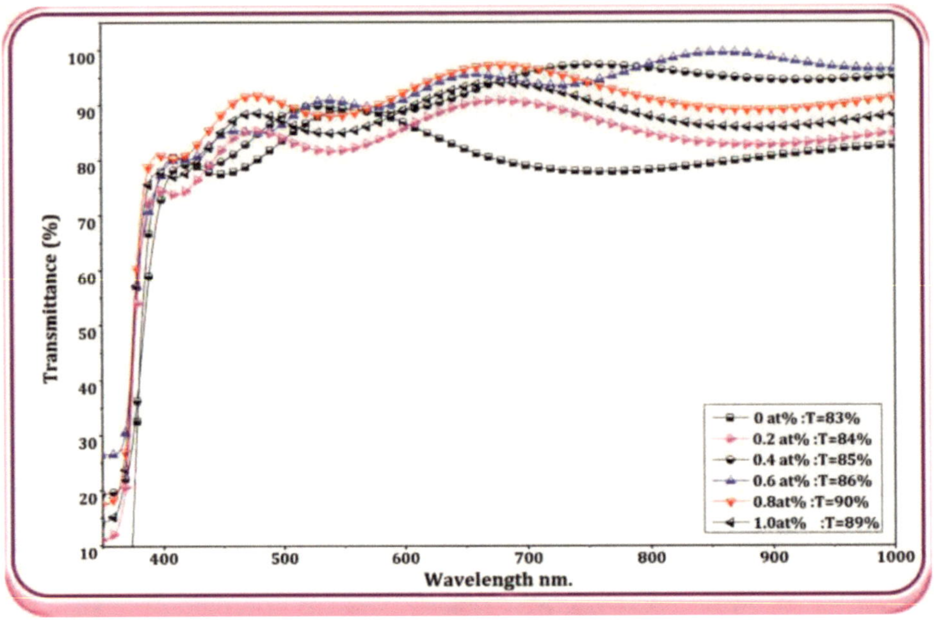

Figure 13. Percentage transmittance of ZnO doped with different % of boron.

5. Conclusion

It has been generally observed that ZnO based thin is very flexible so that it can adapted for some useful applications since the characteristics and properties can easily be modified by doping and annealing. From this review, it was noted that at low temperature, the morphology of ZnO thin film appeared to be smooth as a result of randomly oriented fine-grained polycrystals, but at higher temperature, the smoothness of the morphology become more pronounced with preferred c-axis orientation. Conversely, for ZnO doped with different elements, the grain size in SEM images increased with increase in the percentage concentration of the dopants which is an indication that the dopants influences the physical properties of the film which is invariably enhanced by the annealing. The surface morphology was found generally to be good with the stoichiometric formation of ZnO nanocrystals shape which demonstrates good aggregation of the particles and was suggested to have been originated from the large specific surface area and high surface energy as observed from the structural analysis.. From the XRD analysis in all the cases, ZnO thin film and its doped counterpart annealed at various temperatures depicted high and pronounced intensity at (101), (002) and (100) respectively according to [35, 36] with an increase in peak intensity as annealing temperature is increased in all the cases respectively irrespective of the growth technique and the dopant used during the growth of the film. Optically the spectral absorbance is seen to have high value within the visible widow where the percentage transmittance appeared to have lower value. On the other hand, the transmittance has a high percentage transmittance within near infra-red and infra-red region of electromagnetic wave spectrum. This result then presents ZnO based thin films as good candidate for UV filter and good infra-red transmitter and the thin film being a well-known direct band gap thin film with average band gap of 3.31 eV. From the results it was obviously observed that for both sol–gel and other growth techniques analyzed in this work, there is not much deviation from the values of the band gap. This obviously goes ahead to suggest that ZnO based thin films irrespective of growth technique has a good range of band gap characteristic that made it good for application in optoelectronics, it could also be used as visibly transparent and heating films for use in a cold climate selective windows to transmit only visible and infra-red radiation into buildings while shutting off UV radiation. This will lead to warm indoor temperature in buildings which have their windows coated with such oxide based thin films. The observed wonderful features and tremendous opportunities in ZnO-based heterostructures make it unique and promising in oxide electronics based thin film and has led to new quantum functionalities in optoelectronic devices and electronic applications with lower energy consumption and high performance.

Author details

Emmanuel Ifeanyi Ugwu

Address all correspondence to: ugwuei@yahoo.com

Department of Industrial Physics EBSU, Ebonyi State University, Abakaliki, Nigeria

References

[1] Bunn CW. The lattice-dimensions of zinc oxide. Proceedings of the Physical Society of London. 1950;**475**:835-842

[2] Liu C, Yun F, Markoc H. Ferromagnetism of ZnO and GaN; a review. Journal of Materials Science: Materials in Electronics. 2005;**16**(9):555-597

[3] Tsukazaki A, Ortomo A, Onuma T, Ohitani M, Makine T, Smiya M, et al. Repeated temperature modulation epitaxy for n-type doping and light-emitting diode based on ZnO. Nature Materials. 2005;**4**(1):42-46

[4] Wellings JS, Chaure NB, Heavens NS, Dhamadasa. Growth and characterization of electrodeposited ZnO thin films. Thin Solid Films. 2008;**516**:123893-123898

[5] Rathinama C, Thirukonda AV, Jagannathan T. In: Maruda Y, editor. Growth of Undoped and Metal Doped ZnO Nanostructures by Solution Growth, Nanometal. Croatia: Intech Open Assess; 2010. pp. 31-43. Chapter 2

[6] Kim YS, Tai WP, Shu SJ. Effect of preheating temperature on structural and optical properties of ZnO thin film grown by sol-gel process. Thin Solid Films. 2005;**491**:54-60

[7] Li X, Asher SE, Keyes BM, Luther MJ, Coutts TJ. P-type ZnO thin films grown by MOCVD. In: 31st IEEE Photovoltaic Specialists Conference and Exhibition. 2005. http//www.ost.gov/bridge

[8] Sarma H, Chakrabortty SKC. Optical and structural properties of ZnO thin films fabricated by SILAR method I. Journal of Innovative Research in Science, Engineering and Technology. 2014;**3**(10):16957-16964

[9] Myers MA, Myer MT, General MJ, Lee JH, Shao L, Wang H. P-type ZnO thin film achieved by N^+ ion implantation through dynamic annealing process. Applied Physics Letters. 2012;**101**(11). DOI: 10.1063/1.4751467

[10] Park SH, Monegishi T, Oh DC, Kim DJ, Chang JH, Yoa T, et al. Influence of isoelectronic Te-doping on the physical properties of ZnO films grown by molecular beam epitaxy. Japanese Journal of Applied Physics. 2013;**52**:055501

[11] Joseph B, Manoj PK, Vaidyan VK. Studies on the structural and optical properties of Al-doped ZnO thin films prepared by chemical spray deposition. Ceramics International. 2006;**32**(5):487-493

[12] Oral AY, Batis ZB, Asian MH. Microstructure and optical properties of nanocrystalline ZnO and ZnO (Li or Al) thin films. Applied Surface Science. 2007;**253**(10):4593-4598

[13] Lokhande BJ, Patil PS, Uplane MD. Study on structural, optical and electrical properties of boron doped ZnO thin film prepared by spray pyrolysis technique. Physics B. 2001;**302**:59-63

[14] Srinivasan G, Kumar J. Optical and structural characterization of ZnO thin film prepared by sol gel process. Crystal Research and Technology. 2006;**41**(9):893-896

[15] Tamg ZK, Wang GKL, Yu P, Kawaraki M, Ohtomo A, Koinuma H, et al. Room-temperature ultraviolet laser emission from self-assembled ZnO micro crystallite thin films. Applied Physics Letters. 1998;**72**:3270

[16] Cracium V, Eiders J, Gardeniers JGE, Royd LW. Characteristics of high quality ZnO thin films deposited by pulsed laser deposition. Applied Physics Letters. 1994;**65**:2963

[17] Kotlyarchuk R, Sarchuk V, Oszwaldowski M. Preparation of undoped indium doped ZnO thin films by pulsed laser deposition method. Crystal Research and Technology. 2005;**40**(12):1118-1123

[18] Corta CR, Emanetoglu NW, Liang S, Mago WE, Cu Y, Wraback M, et al. Structural, optical, and surface acoustic wave properties of epitaxial ZnO films grown on $(01\bar{1}2)$ sapphire by metalorganic chemical vapor deposition. Journal of Applied Physics. 1999;**85**:2602

[19] Iribarren A, Fernandez P, Piqueraz J. Cathodoluminescence study of Te-doped ZnO microstructure grown by vapour-solid process. Journal of Materials Science. 2008;**43**:2844

[20] Cho S. Effect of growth temperature on the properties of ZnO thin film grown by radio-frequency magnetron sputtering. Transactions of the Electrical and Electronic Materials. 2009;**10**(6):185-188

[21] Shaki N, Gupta PS. Structural and optical properties of sol-gel prepared ZnO thin films. Journal of Applied Physics Research. 2010;**2**(1):19-28

[22] Ayouchi R, Martin F, Leinen D, Ramos-Barrado JR. Growth of pure ZnO thin films prepared by chemical spray pyrolysis on silicon. Crystal Growth. 2002;**247**:497-504

[23] Chandramohan R, Vijayan TA, Thirumalai J. In: Masuda Y, editor. Growth of Undoped and Metal Doped ZnO Nanostructures by Solution Growth. Nanocrystal. Croatia: Intech Open Access; 2010. pp. 31-43

[24] Znaidi L, Touani T, Vrel JD, Souded N, Yahla SB, Brinza O, et al. ZnO thin films synthesized by sol-gel process for photonic application. Acta Physica Polonica A. 2012;**121**(1): 165-168

[25] Pawar BN, Jadkar SR, Takwale MG. Deposition and characterization of transparent and conductive sprayed ZnO:B thin films. Journal of Physics and Chemistry of Solids. 2005;**66**:1779-1782

[26] Cheng XM, Chien CL. Magnetic Properties of Epitaxial Mn-Doped ZnO Thin Film. 2003

[27] Yadav SC. Synthesis and properties of boron doped ZnO thin films by spray CVD technique at low substrate temperature. International Journal of Environmental Science and Technology (IJEST). 2012;**4**:4893-4898

[28] Burstein E. Anomalous optical absorption limit in InSb. Physics Review. 1954;**93**:632-633

[29] Li M-f. Modern Semiconductor Quantum Physics. Singapore: World Scientific; 2001

[30] Hong D, Gong B, Petella AJ, Russell JJ, Lamb R. Characterisation of the ZnO Thin Film Prepared by Single Course Chemical Vapour Deposition under Low Vacuum Condition, Science in China Series E. 2003

[31] Pogrebnjak AD, Jamil NY, Muhamed AM. Electrical and optical properties of ZnO:Al films prepared by chemical vapour deposition (CVD). In: Proceedings of 1st International Conference "Nanomaterials: Applications & Properties" (NAP-2011). Ukraine, Crimea, Alushta, Sptember 27-30, (2011). 2011

[32] Moss TS. The interpretation of the properties of indium antimonide. Proceedings of the Physical Society. Section B. 1954;**67**:775-782

[33] Grundmann M. The Physics of Semiconductors. 2nd ed. New York, NY: Springer Berlin Heidelberg; 2006

[34] Sönmezoglu S, Akaman E. Improvement of the physical properties of ZnO thin films by tellurium doping. Applied Surface Science. 2014;**318**:319-322

[35] Nguyen HT, Andreas JH, Lamb RN. Structural order of nanocrystalline ZnO film. The Journal of Physical Chemistry. B. 1999;**103**:4264-4268

[36] Oumi Y, Takaba H, Chettu SS, et al. Periodic boundary quantum chemical study on ZnO ultra-violet laser emitting material. Japanese Journal of Applied Physics. 999;**38**(1): 2603-2605

Chapter 3

Sol-Gel Process and Engineering Nanostructure

Habibollah Aminirastabi, Hao Xue,
Dongliang Peng and Gouli Ji

Additional information is available at the end of the chapter

http://dx.doi.org/10.5772/intechopen.79857

Abstract

The production of ceramic nanostructures and engineering of their structure are the goals of the high-tech industry. Researchers prefer the sol-gel route to control the material at the atomic scale among other methods. In this chapter, we describe the production of ceramic nanostructures in different forms such as film, fiber, glass, and powder. We also discuss about microstructures and properties of these ceramic materials and the relationship between them.

Keywords: films, fibers, powder, sol-gel

1. Why engineering nanostructure

Economic necessities, scientific opportunities, and minimization have led to the development of nanotechnology and engineering nanostructure. A key aspiration of nanotechnology is to demonstrate the proposition that as things become small, often, they become differently interesting and useful and valuable structures.

At the atom and molecule scale, behaviors of atoms and molecules are only explicable on the basis of quantum mechanics. For the most part, the requirements of engineering nanostructure will necessitate that these syntheses generate highly homogeneous (both structurally and functionally) nanostructures. The polydispersity that characterizes most syntheses of colloids, for example, without purification, it's impossible to make uniform crystallization [1–5].

2. Microstructure of ceramics

There is a solid connection between the physical properties of ceramics, process, and their microstructure. Subsequently, the significance of the investigation of the microstructure is clear in ceramic research [6]. Materials science is the examination of the relationship among processing, structure, properties, and execution of materials (**Figure 1**).

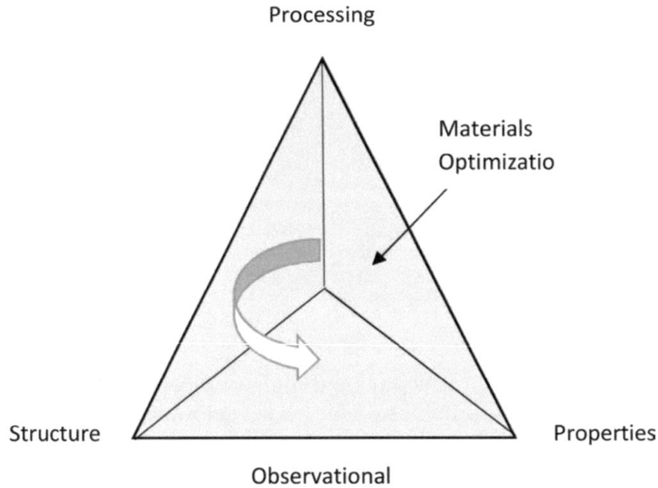

Figure 1. Relationship among processing, structure, and the properties.

3. Evaluation of the microstructure in ceramics

The microstructure and the subsequent properties of ceramics are not static in behavior. They might be changed by outer factors, for example, force and temperature. In the event that the

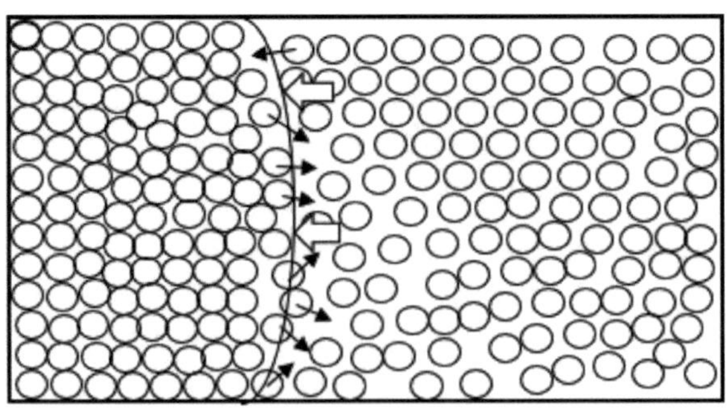

Figure 2. Schematic view of grain growth, showing interdiffusion of the atoms across the boundary. The boundary movement is toward the center of curvature, arising from the stability of the grains.

temperature is high in order to permit huge atom movements, the normal grain size of the ceramics will increase with time. The main impetus for grain growth is the free energy released as the atoms move over the boundary from the convex surface to the concave surface where the atoms end up noticeably planned with the bigger number of neighbors at balance atom spacing. Subsequently, the boundary moves toward the focal point of ebb and flow, and the bigger grains will develop to the detriment of the littler grains, as is shown in **Figure 2**. This is valid in both single-stage microstructures and polyphase microstructures. The net impact is less boundary territory per unit volume [7].

4. Methods of preparation of ceramics in the commercial stage

During the last decade, a wide number of synthesis methods have been developed for the preparation of ceramic powders; the main method is divided into two parts:

4.1. Solid-state reaction method

Industrial production is frequently based on solid-state reactions. But, solid-state reaction processes need high calcination temperature; this requirement leads to many disadvantages of the ceramic powders such as large particle size, wide size distribution, and high degree of particle agglomeration, which generally limits the ability to fabricate reliable electronic components and minimization.

Although this method can be improve with influence mechanochemical process on initial powder and sintering properties but impurities during process milling can be effected to ceramic properties.

4.2. Chemistry-based methods

Different chemistry-based routes have been proposed to produce high-purity ceramic fine powders at low temperatures. These include sol-gel processing, hydrothermal method, spray pyrolysis, oxalate route, microwave heating, microemulsion process, polymeric precursor method, etc. [8].

5. Sol-gel route

The sol-gel processing route was widely studied, because it is very effective in producing ceramic powders with high purity by using a purified precursor. Other advantages of the sol-gel route are good uniformity and a high homogeneous multicomponent system at relatively low temperatures, controllable kinetics of different chemical reactions such as hydrolysis-condensation and nucleation and growth of primary colloidal particles to obtain a microstructure with a special size and size distribution. Also, this route allows to control all of the processes to synthesize tailor-made materials, and homogeneity can be controlled until the atomic scale [9–11].

6. Preparation of ceramics in nanoparticle and fabrication nanosystem

The utilization of nanostructure materials is not another advancement. During the fourth century AD, Romans were utilizing nanosized metal to enhance cups and glasses [12].

Presently, old and novel techniques are utilized to synthesize nanoparticles in an industrial portion. These include sol-gel and chemical wet strategies, flame and spray pyrolysis, and synthetic vapor systems.

As a rule, maximum synthesis techniques and fabrication nanosystem can be divided into two important procedures: "top-down" (microcontact printing and photolithography) and "bottom-up" (self-assembly—organic synthesis) and their combinations. "Top-down" strategies start with a material or a group of macroscopic materials and utilize conventional workshops or microfabrication techniques in which outside controlled apparatuses are utilized for cutting, milling, and shaping materials into the favored form and order, while "bottom-up" approach strategies start with the plan and amalgamation of particles that can self-assemble or self-organize into mesoscale and higher full-scale structures. Bottom-up strategies are utilized to gather atomic and molecular segments into a sorted out surface structure through the natural procedures in the control framework [13].

7. Self-assembly

Self-assembly is one of the phenomena wherein components of all kinds, exemplified by atoms and molecules, colloids, and polymers, have the ability to assemble themselves to form a larger functional unit. It is the key rule that can produce a structure and arrange it from the atomic scale to a huge scale. Finally, with spontaneous organization, a higher level of steady structure is achieved with minimal energy level. Methods and techniques for self-assembly include forces and interactions; some forces used include covalent bonds (chemical bonding), van der Waals interactions, electrostatic interactions, hydrogen bonding, magnetic force bonding, hydrophobic interactions, electrical force, and gravity.

Self-assembly between the atomic scale and the nanoscale with the sol-gel method is possible with control, nucleation, coagulation, and grain growth during the sintering process with control parameters such as temperature, rate of heating, soaking time, and atmosphere of interaction. So, we need to know about some techniques of molecular synthesis, colloid chemistry, and polymer

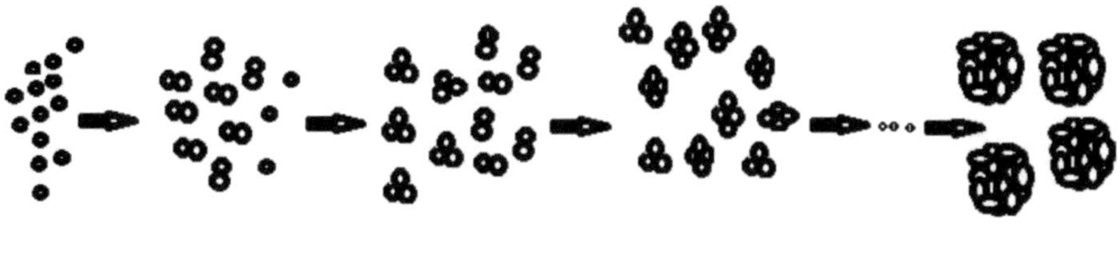

Figure 3. Schematic bottom-up and dynamic assembly of grain during the synthesis process.

science. Self-assembly reflects self-similarity that information coded (such as shape, surface properties, charge, polarizability, magnetic dipole, mass, etc.) in individual components; these characteristics can be determined with fractal dimension and self-similarity according to **Figure 3** [14].

8. Engineering nanostructure with the control grain growth process

Initially, we will define a grain growth concept and then evaluate the grain growth of ceramics, which is a combination of natural growth and self-assembly during growth. We then introduce a new tool that examines this complex phenomenon, which in the future will be able to control the reaction without human intervention and complete intelligence.

9. Grain growth nanostructure

Discovering the connection between the microstructure and macroscopic performance is useful for the improvement and utilization of ceramic materials. Concentrate the grain growth kinetic which has close connection with the assessment of ceramic microstructure. Research on grain growth is of rising interest nowadays and has attracted the attention of several researchers of various disciplines. Nanostructured (NS) materials have a large amount of stored energy due to their large grain boundary area and thus tend to be unstable with respect to grain growth during the sintering process. This problem usually is limited to sintering of NS ceramics at high temperatures. Particularly in industrial applications, how to suppress the grain growth of nanocrystalline materials and how to hold their excellent properties appear to be extremely important [15].

One of the fundamental objectives in sintering nanoparticles is to get dense compacts with held grain sizes in different forms. Hence, understanding subtle elements of the phenomenon of grain growth and the parameters influencing it is indispensable for an effective processing [16].

Ceramics are characterized as an artwork as well as a science of making and utilizing strong articles that have as their basic segment and are made in expansive part out of inorganic non-metallic material. Sintering is an assembling system that has existed for delivering powder metallurgy parts and ceramic segments; amid the standard preparation of ceramics, powders are compacted and after that sintered at a temperature adequate to create helpful properties. Amid the way forward toward sintering, we should be worried about recrystallization and grain growth and, furthermore, the changing size and state of grains and pores [17].

10. Normal grain growth and kinetic exponent

During sintering of nanograins of ceramics, the normal size increases because of coarsening. The established marvel for depicting grain growth is called Ostwald ripening; in other words, expansive grains develop to the detriment of little grains that break up, prompting a stepwise increment of the normal size each time a little grain vanishes. In view of the customary mean-field estimation, the main thrust for grain growth is given by

$$\frac{\Delta G_m}{v_m} = 2\delta_p(\frac{1}{r_c} - \frac{1}{r}) \tag{1}$$

where is the molar volume, the critical radius and the average interfacial energy of the particle, Grains bigger than will develop and littler grains will shrink. Grains having the critical size in this way are in an unstable equilibrium with the matrix.

Ostwald ripening was considered in the traditional investigation exhibited by Lifshitz, Slyozov, and Wagner (the LSW hypothesis). The hypothesis proposes a stationary particle size distribution with a normal size expanding with time. The essential presumption of the LSW hypothesis is that the growth rate is relative to the driving force; in other words, the growth rate diminishes as the grain growth proceeds since the accessible surface energy diminishes amid the procedure.

Boundary movement is impacted by grain size, temperature, and impurities. Smaller grain sizes give a more noteworthy driving force to atom movement across the boundary, which can be spoken to by the following articulation [18]:

$$\frac{dD}{dt} = k/D^m \tag{2}$$

where D is the grain diameter and k and m are the constants. Some trial information demonstrates their qualities that are near solidarity. Accordingly, Eq. (2) disentangles to

$$D^{2-} D_0^2 = 2kt \tag{3}$$

This is a parabolic grain growth law. Thus, the adjustment in the cross-sectional zone of the grain is relative to time, or if the underlying size is thought to be zero, the diameter across increases with the square root of time. Moreover, the estimation of k is typically an exponential capacity of temperature in which k mirrors the activation energy for the atom movements.

11. Abnormal grain growth

Hypotheses of abnormal grain growth (AGG) treat this fascinating marvel regarding the relative grain size, or grain range, of the abnormal grains. Abnormal grain growth is the point at which few of the grains grow all the more definitely contrasting with the encompassing grains amid the sintering procedure. This growth is frequently seen in frameworks having faceted grains. A faceted grain has a particular interface and an anisotropic surface energy, while a circular grain has a harsh interface and an isotropic surface energy. Typically, grain growth of round grains is diffusion controlled and can be portrayed with the LSW hypothesis [19].

Abnormal grain growth dissimilar to normal grain growth is described by the exorbitant development of a moderately modest number of grains, while the rest remain unaltered until the point when they are expended. Because of its temperament, this marvel has been likewise called "grain coarsening," "exaggerated grain growth," "discontinuous grain growth," and furthermore "secondary recrystallization" [20].

12. Classification of nanostructure with dimension

Nanomaterials could have one, two alternately, and three dimensions at the nanoscale.

One-dimensional nanomaterials include layer, multiple layers, thin films, platelets, and also surface coatings. They have been created what's more utilized for decades, especially in the electronic commercial enterprises.

Two-dimensional nanomaterials include nanowires, nanofibers committed starting with an assortment about elements, nanotubes also a subset about this assembly for carbon nanotubes.

Three-dimensional nanomaterials would know as nanoparticles also incorporate precipitates, colloids Furthermore quantum dots (tiny particles from claiming semiconductor materials), and Furthermore Nano-crystalline materials [21].

13. Sol-gel route for production of nanostructure ABO_3

ABO_3 with nanocrystal grains was synthesized by the sol-gel method. For ABO_3 fibers, same layer of ABO_3, acetate of A element, (A $(CH_3COO)_2$), and organic form of element of B same tetrabutyltitanate (B $(OC_4H_9)_4$) for film, fiber or powder were used as main raw reagents, and acetic acid (CH_3COOH) and ethyl alcohol as solvents. **Figure 4** shows the preparation process of nanostructured ABO_3 fibers, film, and powder by the sol-gel route.

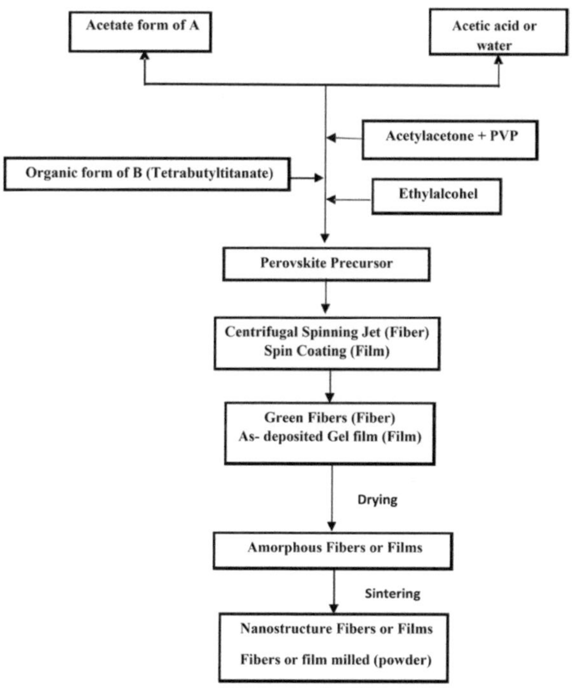

Figure 4. The preparation process of nanostructured ABO_3 fibers, film, and powder by the sol-gel route (Habibollah Aminirastabi Method).

14. Centrifugal spinning with sol-gel solution for production of nanostructure ABO_3 fibers

Different processes are used for synthetizing ceramic nanostructures, such as electrospinning, hydrothermal methods, laser ablation, and chemical vapor deposition (CVD), among which the centrifugal spinning is a highly efficient fiber formation technique that excludes some of the disadvantages of other methods, such as complex processing parameters (e.g., reaction temperature and pressure), low yield, low efficiency, long duration, and costly equipment (e.g., high voltage, reaction chamber, and autoclave) and processes. Hence, this method is capable of increasing the yield, efficiency, and safety in the production process of nanomaterials. **Figure 5a** and **5b** shows fibers sintered at different temperatures [22].

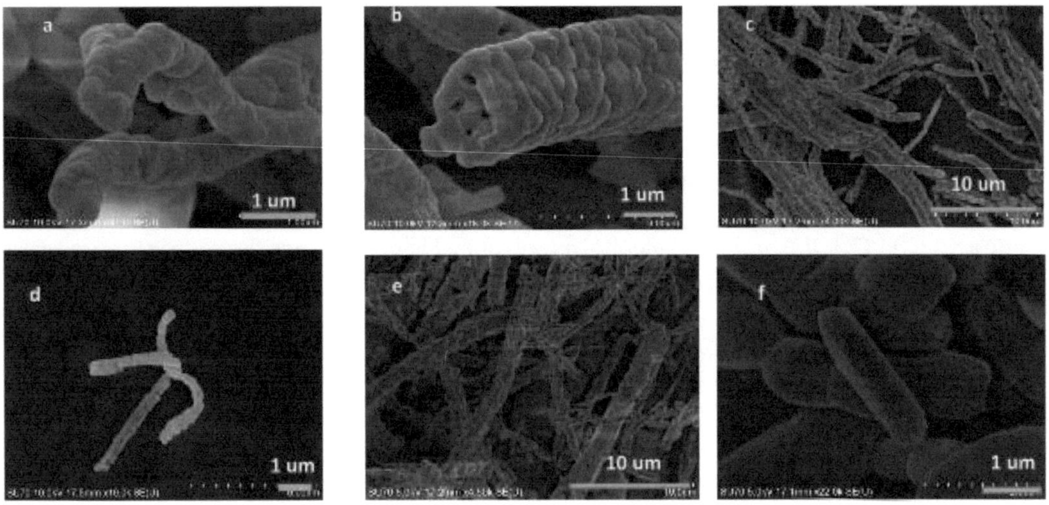

a

b

Figure 5. (a) SEM images of fibers sintered at 1100°C for 30 min: (a and b) $BaTiO_3$, (c and d) $SrTiO_3$, and (e and f) TiO_2. (b) (1–3) SEM micrographs of the $BaTiO_3$ tube sintered at 800^0 C for 2 hours and (4–6) SEM micrographs of TiO_2 rod sintered at 800^0 C for 2 hours.

15. Sol-gel route for production of nanostructure of ABO_3 layer and powder

Organic form of A element such as tetrabutyltitanate (Ti $(OC_4H_9)_4$) and acetate of B element were used as the main raw reagents, and acetic acid (CH_2COOH) and ethyl alcohol were used as solvents. Acetic solution of B element was mixed with polyvinylpyrrolidone dissolved in ethyl alcohol and acetyl acetate, and then tetrabutyltitanate was added. The mixture was stirred vigorously at room temperature for 1 hour to form the ABO_3 precursor. Then, the sol-gel solution was dried and calcined at 700°C for 10 min in different powders as shown in **Figures 6** and **7**.

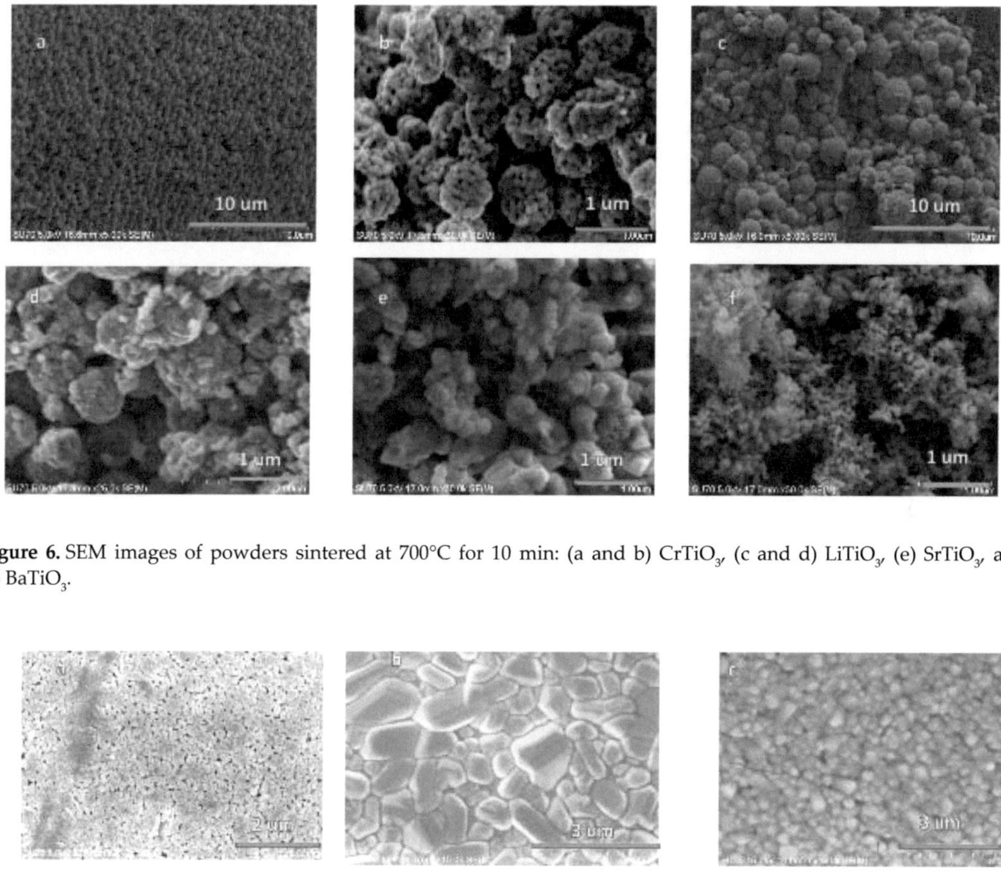

Figure 6. SEM images of powders sintered at 700°C for 10 min: (a and b) $CrTiO_3$, (c and d) $LiTiO_3$, (e) $SrTiO_3$, and (f) $BaTiO_3$.

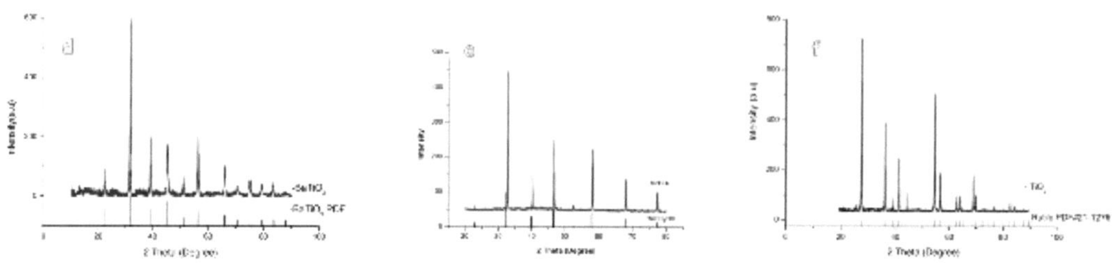

Figure 7. SEM images and XRD patterns of different layers sintered at different temperatures and soaking times: (a and d) $BaTiO_3$ 900°C for 120 min, (b and e) $SrTiO_3$ 1100°C for 120 min, and (c and f) TiO_2 900°C for 120 min.

16. Effect of additives on ABO_3 properties

Normally, ceramics with a perovskite structure has some defects such as point defects, vacancies, interstitial defects, line defects such as edge dislocation and screw dislocation, and plane defects such as grain boundary, tilt boundary, and twin boundary. The crystalline with structural defects that can be corrected by replacing atoms with an atomic radius equal to or smaller, and obtaining new properties with changing microstructure according **Figure 8**. For example, $BaTiO_3$ is doped with small amounts of strontium to improve properties such as permittivity two times more than that of $BaTiO_3$ and to decrease the temperature of sintering.

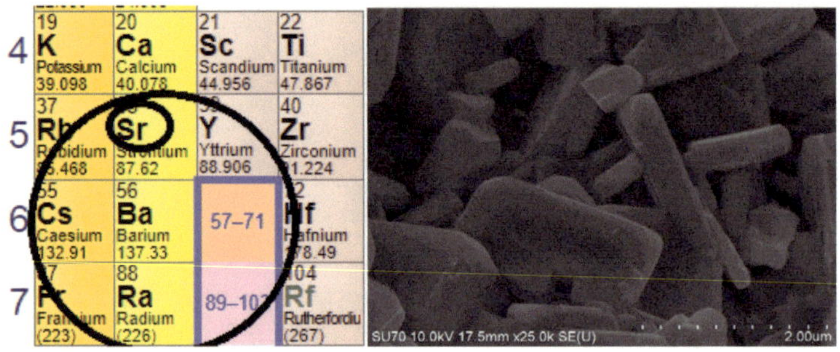

Figure 8. SEM micrograph of $(Ba_{95}Sr_5)$ TiO_3 with microstructure cubic grain.

17. Artificial intelligent tools for the grain growth process

Fractal analysis systems have the ability to make an image of different formats and analysis structures to find and calculate fractal dimension and coverage, from color, gray scale, binary, and 3D sliced (layer) images (**Figure 9**).

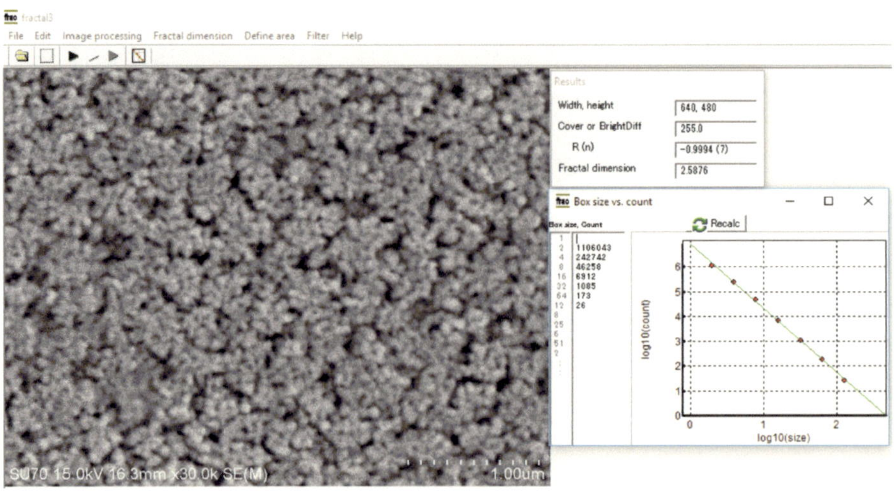

Figure 9. Image analysis of ceramic microstructure of BaTiO3 film sintered at 700°C for 100 minutes.

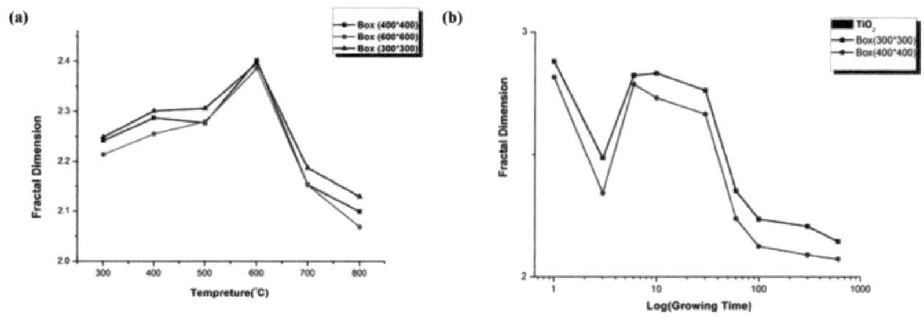

Figure 10. (a) variation of fractal dimension of the nanostructure at different temperatures for 2 hours. TiO_2 sintered at different socking times. (b) Nanostructure TiO_2 sintered at different socking times [24].

Fractal dimension can be calculated by the method of box-counting after preprocessing. The relationship between the size of the box and the count can be displayed with a plot graph. So, you can confirm whether the image is fractal or not by linearity. If limited sizes can be used for calculating fractal dimension, any of the count data can be deleted or edited to recalculate fractal dimension. The image can be filtered; isolated black points can be deleted and filtered for some cases of binary images with much noise. Fractal dimension of black area in many 3D sliced (layer) images (bmp) can be calculated [23] (**Figure 10**).

18. Conclusion

However, there are a number of factors that have influences on the physical properties of ceramics. The production of ceramic nanostructures and engineering of their structure to find single crystal properties are the goals of this research. The advantage of selecting the sol-gel route to produce nanostructures is that the size can be controlled from the atomic scale to the microscale. In this chapter, we describe the production of ceramic nanostructures in different forms such as film, fibers, and powder and some factors that influence grain growth such as temperature, soaking time, and rate of heating and combination of material during the sintering process and introduce tools to monitor the grain growth process, which will be intelligent during all the processes and work based on self-similarity.

Author details

Habibollah Aminirastabi[1]*, Hao Xue[1], Dongliang Peng[1] and Gouli Ji[2]

*Address all correspondence to: habib_amini@yahoo.com

1 Materials College, Xiamen University, Xiamen, Fujian province, China

2 Aerospace Engineering College, Xiamen university, Xiamen, Fujian province, China

References

[1] Hecht DS. Electronic properties of optically transparent single-walled carbon nanotube films. Los Angeles: University of California; 2007

[2] Bernholc J, Brenner D, Nardelli MB, et al. Mechanical and electrical properties of nanotubes. Annual Review of Materials Research. 2003;**8**(32):27-162

[3] Dai J, Tang J, Hsu ST, et al. Magnetic nanostructures and materials in magnetic random access memory. Journal of Nanoscience and Nanotechnology. 2002;**34**(6):281-291

[4] Logothetidis S. Nanostructured Materials and Their Applications. Berlin Heidelberg: Springer; 2012

[5] Wang KL. Issues of nanoelectronics: A possible roadmap. Journal of Nanoscience and Nanotechnology. 2002;**2**(3-4):235-266

[6] Lee WE, Rainforth WM. Ceramic microstructures: Property control by processing. Re fractories. 1994:452-507

[7] Hampshire S. Ceramic processing and sintering. International Materials Reviews. 2003;**41**(1):36-37

[8] Athayde DD, Souza DF, Silva AMA, et al. Review of perovskite ceramic synthesis and membrane preparation methods. Ceramics International. 2016;**42**(6):6555-6571

[9] Chen TD, Wang L, Chen HR, et al. Synthesis and microstructure of boron-doped alumina membranes prepared by the sol–gel method. Materials Letters. 2001;**50**(5):353-357

[10] Chen X, Zhang W, Lin Y, et al. Preparation of high-flux γ-alumina nanofiltration membranes by using a modified sol–gel method. Microporous and Mesoporous Materials. 2015;**214**:195-203

[11] Ahmad AL, Idrus NF, Othman MR. Preparation of perovskite alumina ceramic membrane using sol–gel method. Journal of Membrane Science. 2005;**262**(1):129-137

[12] Mulvaney P. Nanoscience vs. nanotechnology—defining the field. ACS Nano. 2015;**9**(3):2215-2217

[13] Barth JV, Costantini G, Kern K. Engineering atomic and molecular nanostructures at surfaces. Nature. 2005;**437**(7059):671-679

[14] Liu J, Kim AY, Wang LQ, et al. Self-assembly in the synthesis of ceramic materials and composites. Advances in Colloid and Interface Science. 1996;**69**(1-3):131-180

[15] Chen IW, Wang XH. Sintering dense nanocrystalline ceramics without final-stage grain growth. Nature. 2000;**404**(6774):168

[16] Davies AG, Thompson J. Advances in Nanoengineering: Electronics, Materials, Assembly. Imperial College Press; 2007

[17] Logothetidis S. Nanotechnology: Principles and applications. Nanostructured Materials and Their Applications, Springer. 2012:1-22

[18] Beck PA. Effect of recrystallized grain size on grain growth. Journal of Applied Physics. 1948;**19**(5):507-509

[19] Rios PR, Glicksman ME. Topological theory of abnormal grain growth. Acta Materialia. 2006;**54**(19):5313-5321

[20] Cotterill P, Mould PR. Recrystallization and Grain Growth in Metals. Surrey University Press; 1976

[21] Bhatia S. Nanoparticles Types, Classification, Characterization, Fabrication Methods and Drug Delivery Applications. Natural Polymer Drug Delivery Systems. Springer International Publishing; 2016

[22] Rogalski JJ, Bastiaansen CW, Peijs T. Rotary jet spinning review—a potential high yield future for polymer nanofibers. Nanocomposites. 2017;**3**(4):97-121

[23] Xie X. A review of recent advances in surface defect detection using texture analysis techniques. Electronic Letters on Computer Vision and Image Analysis (ELCVIA). 2008;**7**(3):1-22

[24] Aminirastabi H, Weng Z, Xiong Z, Ji G, Xue H. Evaluation Feature of Nano Grain Growth of TiO_2 Thin Film via Sol-Gel Route. In: Materials Processing Fundamentals. Cham: Springer; 2017. pp. 33-43

Chapter 4

Dependence of pH Variation on the Structural, Morphological, and Magnetic Properties of Sol-Gel Synthesized Strontium Ferrite Nanoparticles

Muhammad Syazwan Mustaffa,
Rabaah Syahidah Azis and Sakinah Sulaiman

Additional information is available at the end of the chapter

http://dx.doi.org/10.5772/intechopen.80667

Abstract

In this research work, an attempt of regulating the pH as a sol-gel modification parameter during preparation of $SrFe_{12}O_{19}$ nanoparticles sintered at a low sintering temperature of 900°C has been presented. The relationship of varying pH (pH 1–14) on structural microstructures and magnetic behaviors of $SrFe_{12}O_{19}$ nanoparticles was characterized by X-ray diffraction (XRD), field emission scanning microscope (FESEM), thermogravimetric analysis (TGA), Fourier-transform infrared (FTIR), and vibrating-sample magnetometer (VSM). The single-phase $SrFe_2O_{19}$ with optimum magnetic properties can be obtained at pH 1 with a sintering temperature of 900°C. As pH values increase, the presence of impurity Fe_2O_3 was observed. TGA data-varying pH shows that the total weight loss of most samples was at 30.44% which corresponds to the decomposition process. The IR spectra showed three main absorption bands in the range of 400–600 cm^{-1} corresponding to strontium hexaferrite. SEM micrographs exhibit a circular crystal type of strontium ferrite with an average crystal size in the range of 53–133 nm. A higher saturation magnetization M_s, remanent magnetization M_r, and hysteresis H_c were recorded to have a large loop of 55.094 emu/g, 33.995 emu/g, and 5357.6 Oe, respectively, at pH 11, which make the synthesized materials useful for high-density recording media and permanent magnets.

Keywords: strontium hexaferrite ($SrFe_{12}O_{19}$), sol-gel, pH, structural, magnetic properties

1. Introduction

Ferrite is a magnetic material in the form of ceramic like. Ferrite is usually brittle, hard, iron containing, and generally gray or black in color. It consisted of iron oxides and reacts with preferable high electrical resistivity of metal oxides. Ferrites have impressive properties such as high magnetic permeability and high electrical resistance [1]. Ferrite magnets have a low hysteresis loss and high intrinsic coercivity [2] which give greater effect in resistance demagnetization from external magnetic field. In addition, a low-cost ferrite magnet has good heat resistance and good corrosion resistance which are useful to many applications like permanent magnet [3, 4], solid-state devices, magnetic recording media [5, 6], microwave device [5], etc. A generic formula of magnetoplumbite structure of ferrite is $MFe_{12}O_{19}$, where M is divalent cations like Ba^{2+} [3, 4], Sr^{2+} [1, 2, 5, 7], and Pb^{2+} [8]. Pullar [9] has mentioned that the best known hexaferrite is those containing divalent cations, because it has preferable high electrical resistivity compared to other types of ferrite. $SrFe_{12}O_{19}$ has been chosen in order to produce a good quality of magnetic recording media due to high electrical resistivity of 10^8 Ω cm [9]. The high coercivity leads to high energy product BH_{max} behavior. Liu [10] has mentioned that a good quality of magnetic recording media should have possible high signal and low noise. In order to meet those criteria, the magnetic materials should have high magnetization; high coercivity but correlated with recording field; single-domain particles or grains; a smaller size of particles or grain size, thermally stable, and therefore a reduced thickness of the active magnetic film of the medium; and a good alignment of the particle or grain easy axis [10]. In recent years, higher levels of recording density have been achieved in the field of magnetic recording. Magnetic tapes employing hexagonal barium ferrite magnetic powder achieve a surface recording density of 29.5 bpsi (bits per square inch). However, when the size of hexagonal ferrite magnetic particles is reduced, the energy for maintaining the direction of magnetization of the magnetic particles (the magnetic energy) tends to become inadequate to counter thermal energy, and thermal fluctuation ends up compromising the retention of recording.

Various techniques are presented for the synthesis of strontium hexaferrite powders such as solid-state synthesis method [11, 12], chemical coprecipitation [13–15], ceramic method [16], and sol-gel [17–19] and hydrothermal methods [20]. The effect of pH variation in this research work via sol-gel method for producing $SrFe_{12}O_{19}$ is key factor for controlling hexaferrite nanostructure and magnetic properties. Other than that, this proposed method has not yet been reported elsewhere in producing $SrFe_{12}O_{19}$ nanoparticles. Recently, the sol-gel route has received considerable attention in the last few years because it has lower calcination temperature, the fact that it also enables smaller crystallites to grow [2]. Sol-gel method produces a better outcome than microemulsion and coprecipitation methods. The sol-gel hydrothermal method combines the advantages of the sol-gel method and the high pressure in the hydrothermal condition [7]. In the hydrothermal process, the particle size and particle morphology can be controlled. $SrFe_{12}O_{19}$ nanoparticles have high purity, ultrafine size, and high coercivity. Some efforts have been carried out to modify the sol-gel process parameters such as pH, basic agent, carboxylic acid, and starting metal salts for further decreasing the calcination temperature and achieving the finer crystallite size [1]. Optimizing the molar ratio of Fe to Sr is very important to produce a single-phase sample, ultrafine particle, and lower calcination temperatures [21]. This ratio varies with the change in starting materials and with the change in method of production [21]. The obtained products that have single-phase particles have a hexagonal shape, the right proportion, and high coercively. The prolonging annealing time has a significant effect on the high saturation magnetization (M_s), and the high annealing rate formed a highly percentage of pure strontium hexaferrite. Masoudpanah and Ebrahimi [2] state that the preferred molar ratio of Fe/Sr is 10, which is the lowest calcination temperature (800°C) on the formation of single phase of SrM thin films. In addition, XRD showed that the crystallite sizes at a range of 20–50 nm. The magnetic properties of this preferred molar ratio exhibit a good saturation magnetization (267 emu/cm^3), high coercivity (4290 Oe), and a relatively high remanent magnetization (134 emu/cm^3). Minh et al. [7] state that the preferred molar ratio is at 11. The obtained $SrFe_{12}O_{19}$ has high purity, ultrafine size, and high coercivity at H_c = 6315 Oe. This chapter discussed an attempt to employ water as the gel precursor to synthesize nano-sized M-type strontium ferrite ($SrFe_{12}O_{19}$) bulk sample at low sintering temperature 900°C by using a common laboratory chemical. A solution of metal nitrates and citric acid and ammonia has been used to prepare strontium hexaferrite at varying pH.

2. Brief overview of preparation methods

2.1. Raw materials

Strontium nitrate anhydrous granular Sr $(NO_3)_2$ (98%, Alfa Aesar), iron (III) nitrate $Fe(NO_3)_3$ (99%, HmbG), citric acid (CA) $C_6H_8O_7$ (99%, Alfa Aesar), ammonia NH_4OH (25%, SYSTERM), and deionized water were used as starting material in order to synthesize $SrFe_{12}O_{19}$ nanoparticles as listed in **Table 1**.

The general chemical equation for desired $SrFe_{12}O_{19}$ samples is weighed according to the formula:

$$Sr(NO_3)_2 + 12Fe(NO_3)_3 \xrightarrow[NH_4OH]{C_6H_8O_7} SrFe_{12}O_{19} + volatile \quad (1)$$

The nitrates were calculated as one mole of $Sr(NO_3)_3$, and 12 moles of $Fe(NO_3)_2$ were needed in order to synthesize one mole of $SrFe_{12}O_{19}$ nanoparticles. In the process of reaction, CA was used as a chelating agent and fuel of combustion. The CA was then calculated according to the molar ratio of citrate to nitrate of 0.75 which first obtained each number of mole nitrate as below:

$$Sr(NO_3)_2 \rightarrow Sr^{2+} + 2(NO_3)^- \quad (2)$$

$$Fe(NO_3)_3 \rightarrow Fe^{3+} + 3(NO_3)^- \quad (3)$$

Then, mass of citrate was calculated as:

$$Mass_{CA} = (0.75 \times total\ nitrate) \times molar\ mass_{CA} \quad (4)$$

In this study, NH_4OH was used to vary the pH value of $SrFe_{12}O_{19}$ in order to study the effect of pH value in its morphology and magnetic properties.

Chemical name	Compound formula	Molecular weight (g/mol)	Weight ratio (g)
Strontium nitrate anhydrous (salt)	$Sr(NO_3)_2$	211.63	0.4183
Iron (III) nitrate	$Fe(NO_3)_3$	403.84	9.5817
Citric acid (powder)	$C_6H_8O_7$	191.12	11.8368
Ammonia	NH_4OH	35.04	Varied depend on pH Measured by pH meter

Table 1. Compound used for sol-gel synthesis.

2.2. Sample preparation and characterizations

An appropriate amount of Sr $(NO_3)_2$, $Fe(NO_3)_3$, and $C_6H_8O_7$ was dissolved in 100 ml of deionized water for 30 min at 50°C with constant stirrer rotation of 250 rpm. The mixtures were continuously stirred, and NH_4OH was added in order to vary the pH from pH 1–14 which is measured by HI 2211 pH/ORP meter (HANNA instruments). The solutions then were stirred on the hot plate for 24 h at 60°C. The solution was left in oven at temperature of 80°C for 2 days to turn the solution into a sticky gel. The sticky gel was stirred again stirred on hot plate, and the temperature was increased up to 150°C to dehydrate and form a powder. The powder formed were crushed by using mortar before sintering it at 900°C for 6 h with the heating rate of 3.5°C/min. The crystalline structural characterization of XRD was performed using a Philips X'Pert X-ray diffractometer model 7602 EA Almelo with Cu Kα radiation at 1.5418 Å. The range of diffraction angle used is from 20 to 80° at room temperature. The accelerating current and working voltage were 35 mA and 4.0 kV, respectively. The data are then analyzed by using X'Pert Highscore Plus software. The lattice constant, *a*, is obtained by Eq. (5):

$$a = d\sqrt{h^2 + k^2 + l^2} \tag{5}$$

Where *d* is the interatomic spacing and (*h k l*) are miller indices. The volume cell V_{cell} was calculated using Eq. (6):

$$V_{cell} = \frac{\sqrt{3}}{2} a^2 c \tag{6}$$

Where *a* and *c* are lattice constants. The theoretical density ρ_{theory} of sample was calculated using Eq. (7):

$$\rho_{\text{theory}} = \frac{2M}{N_A V} \tag{7}$$

Where *M* is molecular weight of $SrFe_{12}O_{19}$, which is equal to 1061.765 g. The weight of two molecules in one unit cell is 2 × 1061.765 = 2123.53 g; N_A is the Avogadro's number (6.022×10^{23} mol^{-1}). The porosity *P* of the samples can be calculated using Eq. (8):

$$P = \left(\frac{1 - \rho_{\exp}}{\rho_{\text{theory}}} \right) \times 100\% \tag{8}$$

Where $\rho_{\exp}$ is the experimental density and ρ_{theory} is the xrd density.

Meanwhile, the crystallite size can be measured by using the Scherrer equation (Eq. 9):

$$D = \frac{k\lambda}{\beta \cos\theta} \tag{9}$$

Where D is crystallite size, k is the Scherrer constant value of 0.94, λ is Cu Kα radiation wavelength of 1.542 Å, β is half-peak width of diffraction band, and θ is the Bragg angle corresponding to the planes.

The thermal stability of these samples was obtained by using TGA/SDTA 851 of Mettler Toledo thermogravimetric analyzer. The sample weighted about 10 mg was used at operating temperature range from 0 to 1000°C with heating rate 5°C/min. Fourier-transform infrared by Perkin Elmer model 1650 was used to determine the infrared spectrum of absorption and emission bands of sample. It was performed between infrared spectra of 280–4000 cm^{-1} with resolution of 4 cm^{-1}. The micrograph of microstructure was observed using a FEI Nova NanoSEM 230 machine to study the morphology and microstructure of solid material. The sample was prepared in bulk pallet at a diameter of 1 cm and coated with gold in order to avoid charge buildup as the electron beams are scanned over the samples' surface. The distribution of grain size image was fixed at magnification of 100,000X with 5.0 kV. The distribution of average grain size of microstructure was calculated by using these images. The distributions of grain sizes were obtained by taking at least 200 different grain images for the sample and estimating the mean diameters of individual grains by using the J-image software. The magnetic properties of samples were measured by VSM Model 7404 LakeShore. The measurement was carried out in the room temperature with sample weight about 0.2 g. The external field applied was 12 kOe parallel to the sample. From this analysis, saturation magnetization, M_s; remanent magnetization, M_r; and coercivity, H_c, were recorded, and the hysteresis loop was plotted.

3. Research findings and outcomes

3.1. Structural analysis

Figure 1 shows the XRD spectra of the samples sintered at 900°C with different pH values (pH 1–14). The XRD spectrum shows the formation of a single phase of $SrFe_{12}O_{19}$ nanoparticles. The structure of XRD peaks was referred to standard strontium hexaferrite ($SrFe_{12}O_{19}$) with JCPDS reference code of 98002-9041 [22], with hexagonal crystal system belonging to space group of *P63/mmc* that proved the hexagonal crystal structure system formation. The $SrFe_{12}O_{19}$ phase formed with miller indices shown as [110], [008], [017], [114], [021], [018], [023], [116],

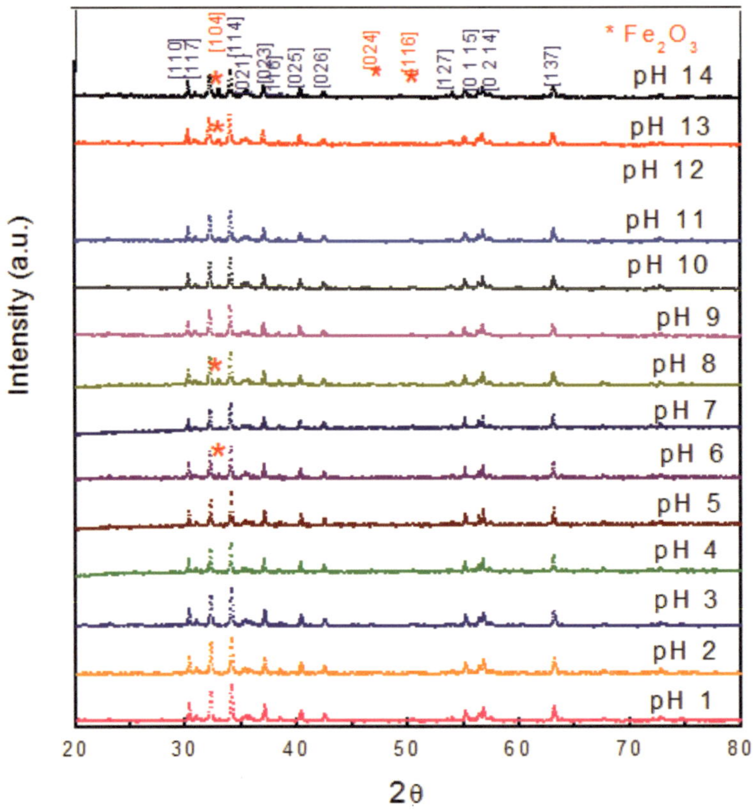

Figure 1. The X-ray diffraction spectra of $SrFe_{12}O_{19}$ nanoparticles for pH 1–14 sintered at 900°C.

[025], [026], [127], [034], [0211], [0115], [0214], and [137], respectively. The highest intensity can be observed at 2θ (34.218°) with miller indices of [114] of reference index code of 98-002-9041 [22]. However, as pH increases, the amount of ammonia required increases. This factor leads to the formation of hematite (Fe_2O_3) in pH 6, pH 8, pH 13, and pH 14 due to excess ammonia that could not completely vanish during reaction. The formation of Fe_2O_3 occurred at 2θ = 33.139 and 49.673° with miller indices of [104] and [024]. The hematite Fe_2O_3 patterns were indexed to ICSD reference code of 98-005-3678 [23]. It was explained by Masoudpanah and Ebrahimi [2] that the increasing pH of the sol results in the absorption of positively charged Sr ions on iron gels and the formation of negatively charged iron gels. A single-phase $SrFe_{12}O_{19}$ was obtained at a low sintering temperature of 900°C for powder pH which proves the benefit of using sol-gel method in this $SrFe_{12}O_{19}$ reaction. It was agreed that obtained single phase of $SrFe_{12}O_{19}$ at lower temperature was due to the solubility of Sr $(NO_3)_2$ that decreases at elevated temperatures [24]. Hence, more Sr^{2+} ions are needed for the formation of the strontium hexaferrite [2]. The diffusion rates increased in the nonstoichiometric mixtures because of the induced lattice defects which could be observed from lower lattice parameter [2].

The average crystallite size (**Table 2**) determined from the full width at the half maximum (FWHM) of the XRD patterns was calculated using the Scherrer formula provided from X'Pert

pH	Peak pos. 2θ (°)	Miller indices (hkl)	Peak width (°)	Space group	Lattice constant		V_{cell} (nm^3)	ρ_{xrd} (g cm^{-3})	ρ_{exp} (g cm^{-3})	*P* (%)	Calculated crystalline size, *D* (nm)
					a (Å)	c (Å)					
1	34.20	[114]	0.13	*P63/mmc*	5.883	23.018	5.11	4.634	13.899	0.690	63.226
2	34.21	[114]	0.13	*P63/mmc*	5.882	23.051	5.11	4.399	11.217	0.691	63.228
3	34.22	[114]	0.16	*P63/mmc*	5.882	23.051	5.11	3.832	8.077	0.691	51.372
4	34.20	[114]	0.16	*P63/mmc*	5.884	23.058	5.10	4.693	13.237	0.691	51.369
5	34.25	[114]	0.16	*P63/mmc*	5.880	23.040	5.11	4.200	12.114	0.690	51.376
6	34.18	[114]	0.16	*P63/mmc*	5.884	23.060	5.10	4.492	11.831	0.691	51.366
7	34.18	[114]	0.18	*P63/mmc*	5.884	23.057	5.10	4.497	12.368	0.691	45.661
8	34.17	[114]	0.18	*P63/mmc*	5.885	23.058	5.10	3.419	9.633	0.691	45.660
9	34.12	[114]	0.14	*P63/mmc*	5.889	23.025	5.10	4.633	9.153	0.691	57.266
10	34.19	[114]	0.15	*P63/mmc*	5.884	23.047	5.10	4.784	6.254	0.691	54.685
11	34.18	[114]	0.15	*P63/mmc*	5.885	23.053	5.10	4.721	7.435	0.691	53.231
12	34.22	[114]	0.17	*P63/mmc*	5.880	23.030	5.12	4.699	8.125	0.689	47.762
13	34.14	[114]	0.18	*P63/mmc*	5.884	23.066	5.10	4.705	7.736	0.691	46.693
14	34.20	[114]	0.18	*P63/mmc*	5.882	23.023	5.11	4.612	9.782	0.690	46.941

Table 2. The summary of the of $SrFe_{12}O_{19}$ nanoparticles for pH 1–14.

Highscore Plus software. From the plotted crystallite size relationship with pH of samples (**Figure 2**), it shows two groups of crystallite size distribution: acidic group (1) for pH 1–8 and alkaline group (2) for pH 9–14. Both groups show an improvement of crystallinity that gives out smaller crystallite size as the pH increases, which results in smaller grain size as the crystallite size increases.

The lattice constant *a* and *c* values (**Table 2**) observed were not far different from the theoretical $SrFe_{12}O_{19}$ lattice constant, where a = 5.8820 Å and c = 23.0230 Å [25], as similar as reported by

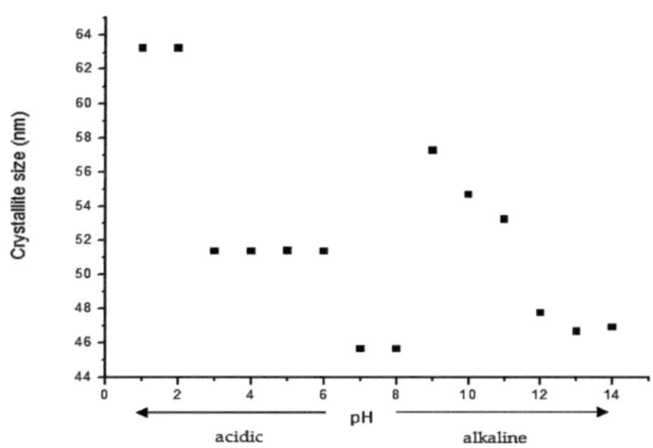

Figure 2. Relationship of crystallite size versus pH.

Masoudpanah et al. [2, 26] and Dang et al. [27]. There is a slight increment in lattice constant *c* as pH increases and fluctuated data of lattice constant *a*. It is shown that, at pH 10, the lattice constant *a* of 5.884 Å was the highest peak with a lower peak of lattice constant *c* of 23.047 Å. The standard strontium hexaferrite ($SrFe_{12}O_{19}$) with JCPDS reference code of 98-002-9041 [22] has theoretical density of 5.11 g cm^{-3} [25]. Theoretically, the density of the sample, ρ_{EXP}, is affected by the lattice constants *a* and *c*. The lattice parameter *a* and *c* values observed were not far different from the theoretical $SrFe_{12}O_{19}$ lattice constant, where a = 5.8820 Å and c = 23.0230 Å (**Figure 3**) [25]. The *a* and *c* parameters observed are similar to Masoudpanah et al. [2] and Dang et al. [27].

The lattice constant was fluctuated around the theoretical lattice constant. However, in the experiment, the density was more affected by the preparation of the sample which results in porosity of the sample. The distant the difference of density of XRD (ρ_{XRD}) and experimental density (ρ_{EXP}), the higher the number of porosity, which results in reducing the mass of the pallet sample by pores. The highest density value for ρ_{XRD} is at pH 12 (5.1148 gcm^{-3}), and the highest density value for ρ_{EXP} is at pH 10 (4.784 gcm^{-3}). The porosity occurs because of the presence of pores in the samples as a result after sintering of bulk samples. The pores occur due to an error from preparing sample and the loosen powder while pressing the sample using hydraulic presser. As the ρ_{EXP} approaches to the ρ_{XRD}, the pores' percentage becomes lower. The highest porosity of 13.24% was found at pH 4 with ρ_{XRD} of 5.1001 g cm^{-3} and ρ_{EXP} of 4.425 g cm^{-3}. Meanwhile, pH 10 exhibits a lower porosity of 6.254% with ρ_{XRD} of 5.1032 g cm^{-3} and ρ_{EXP} of 4.784 g cm^{-3} (**Table 2**).

The powder was synthesized using a control molar ratio of 1:12 with respect to strontium and nitrate. However, the sample with various pH was prepared with an addition of nitrate into the solution. The sample ratio of Sr $(NO_3)_3$ and $Fe(NO_3)_2$ is (Fe/Sr) = 12:1, and the samples were sintered at 900°C. From previous work reported, a single phase of strontium ferrite ($SrFe_{12}O_{19}$) was obtained for samples sintered at 850°C with Fe/Sr molar ratio of 11.5 via sol-gel route [28].

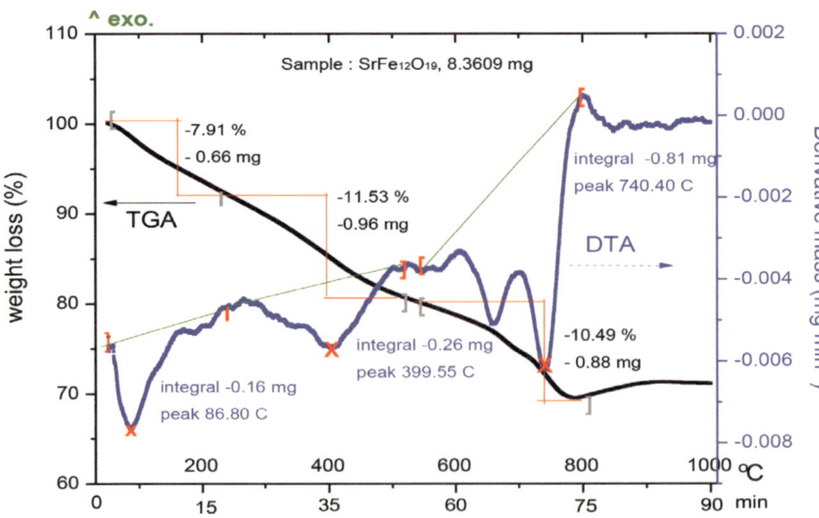

Figure 3. The TGA and DTA traces for dried powder of $SrFe_{12}O_{19}$ sintered up to 1000°C.

Masoudpanah and Ebrahimi [2] found a single phase of $SrFe_{12}O_{19}$ at sintering temperature of 900°C prepared using sol-gel technique. In general, the lowest sintering temperature of $SrFe_{12}O_{19}$ is around 800–1000°C. Hence, the raw powder (non-sintered) was tested by TGA to identify the best temperature by sintering up to 1000°C. The TGA curves as plotted in **Figure 3** show a decreasing amount of weight as the powder sintered up to 1000°C in 20 min with a starting weight of 8.3609 mg. Meanwhile, the DTA diagrams reveal three peaks shown at range 86.80–100, 399.55, and 740.40°C due to decomposition process. At a constant heating rate, the endothermic peak at 86.80–100°C had −7.91% of weight loss due to the dehydration of the absorbed water as the powder slowly turns into burnt gel [2]. The first exothermic peak at 399.55°C with a weight loss of −11.53% is due to the elimination of the organic compound which tends to the decomposition of NH_4NO_3 that liberates NO, O_2, and H_2O [2]. Meanwhile, at stage 740.40°C, the exothermic peak with a weight loss of −10.49% shows the decomposition of citric acid and the breakdown of the Fe_2O_3 to Fe as reported [17]. The stable temperature is at 880°C which permits the completeness of reaction. Hence, sintering temperature at 900°C was used in this work.

Figure 4 shows the FTIR spectra of $SrFe_{12}O_{19}$ nanoparticles for pH variation (pH 1–14), with IR range of 400–4000 cm^{-1}. It is noticeable that spectrum appeared in the range of 430, 583, 904, and 1446 cm^{-1} of IR characteristic band. The stretching band of CH_2 appeared at 436 cm^{-1} attributed to the presence of CH saturated compound, which has been agreed by [29]. The vibration of CH bond could be due to the chemical reaction in a process of hexagonal structure form, where the CH bond of citric acid loses their CH bond. The spectrum of metal-oxygen vibration of Sr–O Fe–O was found at 583 cm^{-1} [26]. Masoudpanah and Ebrahimi [2] explained that an occurred reaction between citric acid and ferric ions is attributed to the stretching mode

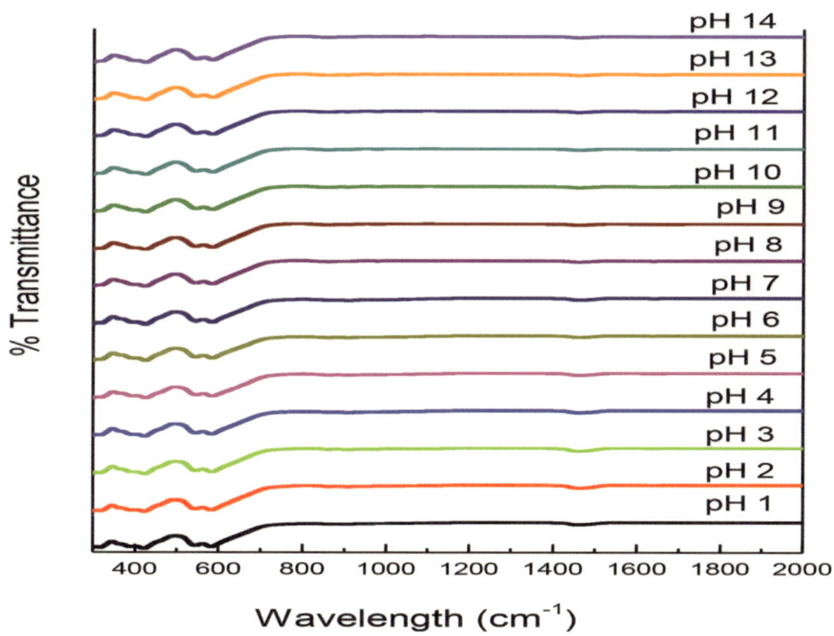

Figure 4. FTIR spectra of $SrFe_{12}O_{19}$ for pH variation sintered at 900°C.

of Fe–O, which confirms the formation of chelate in sol-gel route. It is proven by many researchers who claim that the absorption bands at range 443–600 cm^{-1} were results of the formation of strontium ferrite as the stretching vibration of metal-oxygen bond of Sr–O Fe–O occurs [30–33]. All pH reveals these two bond bands. However, there were some reducing and vanished bands in the next bond bands at 904 and 1460 cm^{-1}. It is due to the purity of $SrFe_{12}O_{19}$ nanoparticles, as there was some interruption of Fe_2O_3 in the sample as shown in **Figure 1** and **Table 2**. In this study, pH 8, pH 10, pH 13, and pH 14 come out with a percentage of hematite, Fe_2O_3. First, the pure $SrFe_{12}O_{19}$ (pH 1–7, pH 9, pH 11–12) had a relatively strong and broad bands at peak 904 cm^{-1}, which revealed that there was amine functional group for N–H vibration due to decomposition of NH_3. Pereira et al. [32] stated that this broad vibration of Sr–O stretching indicates the formation of strontium nanoferrites. It is agreed by Sivakumar et al. [34] that the strontium ferrite was formed and the iron oxide vanished at 900 cm^{-1}. Meanwhile, pH 6, pH 8, pH 13, and pH 14 show a relative small vibration band at 904 cm^{-1} due to the presence of Fe_2O_3. As pH increases up to pH 14, the amount of ammonia increases gradually. Excess amount of ammonia failed to completely decompose the NH_3 bonds and break down the N–H vibration. Lastly, the absorption bands at 1446 cm^{-1} found in pure pH sample were attributed to the vibrating bands of Fe-O-Fe due to the decomposition of metal with oxide band [29]. There was some significant data that show in pH 9, pH 11, and pH 12, as a single phase of samples is formed as pH increased.

3.2. Microstructural analysis

The microstructure and grain size distribution of bulk $SrFe_{12}O_{19}$ nanoparticles are shown in **Figure 5**. The grain size seems to have agglomerated and charged nanoparticles when increasing the pH value. The grain size was found in the range of 53.22–133.25 nm. The pH 4 produces pores of 13.24%. Meanwhile, the most packed grains are for sample at pH 10, with porosity of 6.25% (**Table 2**). The microstructure shows that some of the samples have a large porosity due to the presence of polyvinyl alcohol during the preparation of pellet bulk $SrFe_{12}O_{19}$ nanoferrites. The histogram of the grain distribution was shifted from small grain sizes to exhibiting larger grains from pH 1 to 8. Nevertheless, the grain size was observed to be decreasing as the pH is reaching 9–14 (**Table 3**).

3.3. Magnetic behaviors

The development of *M-H* hysteresis loop at various pH is illustrated in **Figure 6**. The magnetic saturation, M_s; remanent magnetization, M_r; coercivity, H_c; grain size; and porosity of $SrFe_{12}O_{19}$ nanopowder are shown in **Table 4**. An obvious erect, larger, and well-defined hysteresis loop can be observed. It is probably due to the strong ferromagnetic behavior, indicating the formation of $SrFe_{12}O_{19}$ nanoparticles with high volume fraction of the complete crystalline $SrFe_{12}O_{19}$ phase. Thus a strong interaction of magnetic moments within domains occurred due to exchange forces. This observed phenomenon can be considered as ordered magnetism in the sample. In fact, in order to obtain an ordered magnetism and well-formed

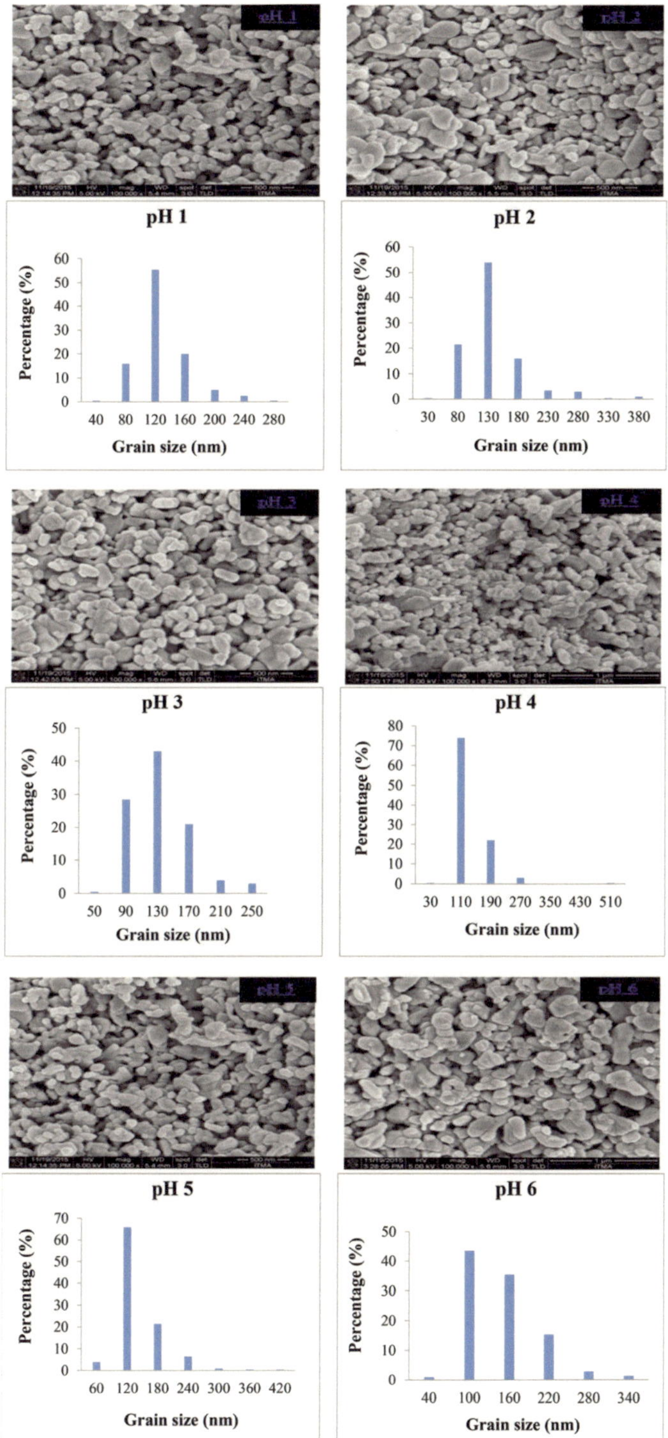
pH 1
Percentage (%)
0 10 20 30 40 50 60
40 80 120 160 200 240 280
Grain size (nm)
pH 2
Percentage (%)
0 10 20 30 40 50 60
30 80 130 180 230 280 330 380
Grain size (nm)
pH 3
Percentage (%)
0 10 20 30 40 50
50 90 130 170 210 250
Grain size (nm)
pH 4
Percentage (%)
0 10 20 30 40 50 60 70 80
30 110 190 270 350 430 510
Grain size (nm)
pH 5
Percentage (%)
0 10 20 30 40 50 60 70
60 120 180 240 300 360 420
Grain size (nm)
pH 6
Percentage (%)
0 10 20 30 40 50
40 100 160 220 280 340
Grain size (nm)

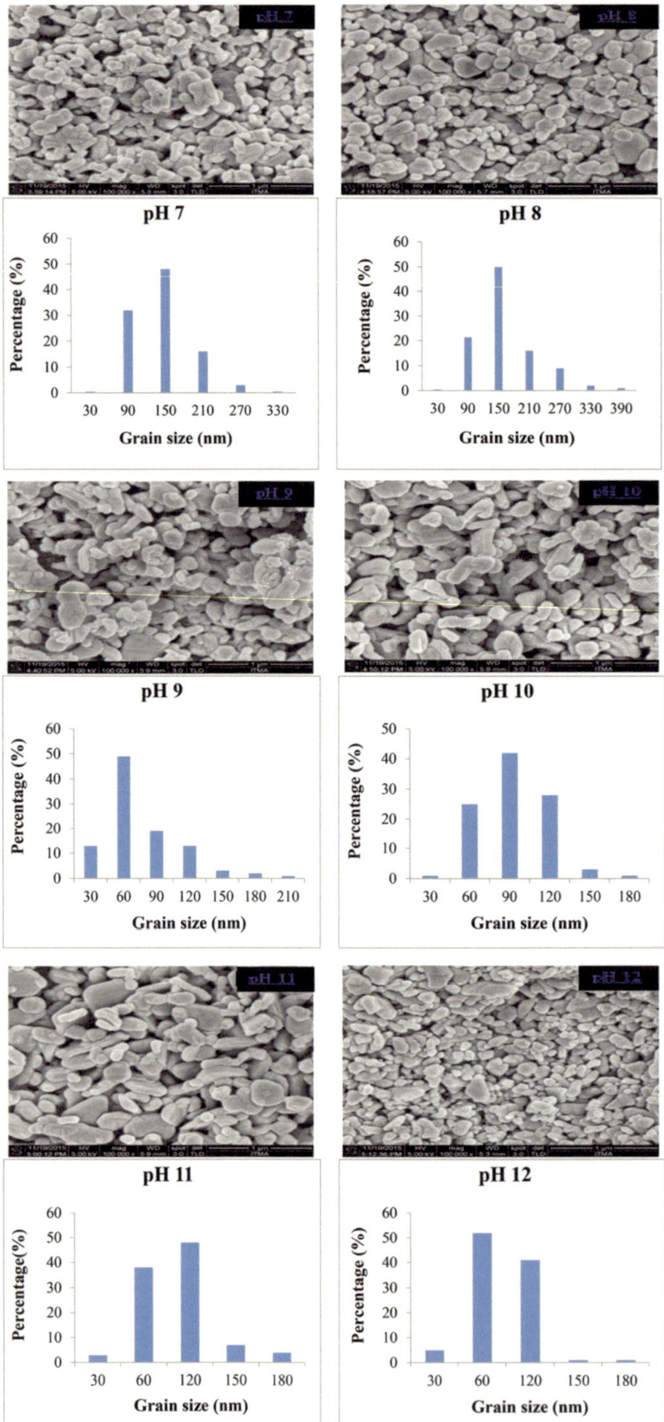
pH 7
Percentage (%)
Grain size (nm)
30 90 150 210 270 330
pH 8
Percentage (%)
Grain size (nm)
30 90 150 210 270 330 390
pH 9
Percentage (%)
Grain size (nm)
30 60 90 120 150 180 210
pH 10
Percentage (%)
Grain size (nm)
30 60 90 120 150 180
pH 11
Percentage(%)
Grain size (nm)
30 60 120 150 180
pH 12
Percentage (%)
Grain size (nm)
30 60 120 150 180

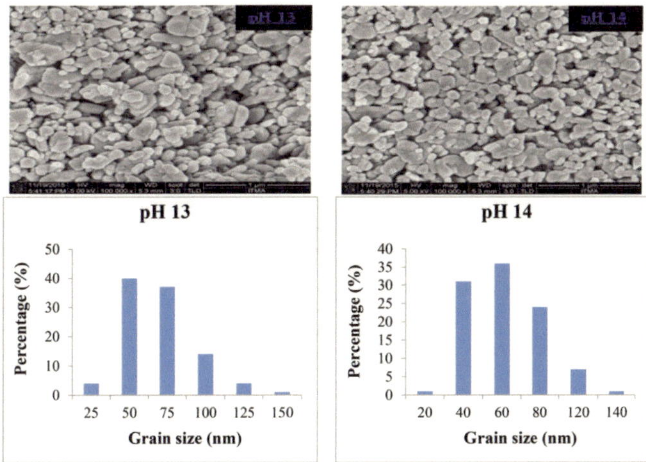

Figure 5. The micrograph image and grain size distribution of $SrFe_{12}O_{19}$ sintered at 900°C by varying pH value.

pH	Grain size (nm)
1	108
2	114
3	115
4	96
5	111
6	120
7	116
8	133
9	61
10	79
11	75
12	62
13	57
14	53

Table 3. Grain size of $SrFe_{12}O_{19}$ sintered at 900°C by varying pH value.

M-H hysteresis loop, there must exist a significant domain formation, a sufficiently strong anisotropy field (H_a), and optional addition contributions, which come from defects such as grain boundaries and pores [35]. The saturation magnetization (M_s), remnant magnetization (M_r), and coercivity (H_c) are found to decrease with increasing pH by addition of ammonia in the sol-gel precursor. From the previous study, the H_c is 4290 Oe, obtained at pH 7 [2, 26]. The

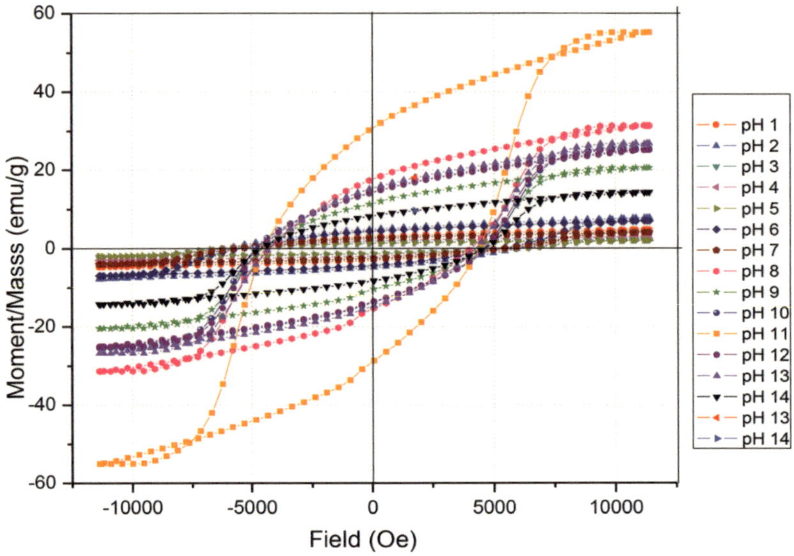

Figure 6. The *M-H* hysteresis loop $SrFe_{12}O_{19}$ of pH 1–14 sintered at 900°C.

pH	Saturation magnetization, M_s (emu/g)	Remnant magnetization, M_r (emu/g)	Coercivity, H_c (Gs)
1	4.776	3.001	6094.7
2	7.822	4.870	6005.8
3	2.168	1.373	5966.1
4	3.006	1.929	5808.6
5	2.016	1.309	6074.8
6	7.022	4.416	5377.0
7	4.028	2.554	5461.2
8	31.342	19.363	5058.3
9	20.488	12.776	5422.2
10	25.471	15.825	5663.1
11	55.094	33.995	5357.6
12	25.114	15.674	5532.7
13	26.849	16.885	5185.9
14	14.239	9.1325	5520.7

Table 4. M_s, M_r, and H_c of $SrFe_{12}O_{19}$ as a function of pH.

microstructures of nanoparticles were affected by the increase of pH value. This is in agreement with findings reported by Yang et al. [36], where the formation of particles became larger [37] with the increase of pH from 5 to 11. This is due to the aggregation of small particle that occurs when there is a strong magnetic interaction between magnetic atoms (Fe or Co) containing in Co-Fe-Al grains as the composition of Co increases and the composition of Al decreases [38]. In this work, it is noticeable that pH 11 has the largest hysteresis loops as well as high magnetic properties. Moreover, the remaining pH exhibit almost the same hysteresis loop with a slight change in M_s and M_r. Meanwhile, the presence of Fe_2O_3 impurity in the samples of pH 6, 8, 13, and 14 shows a decrease in H_c, which affects the crystalline and grain boundary. The H_c is observed to reduce as pH increased. The presence of intragranular trapped pores in the grains was due to rapid grain growth of sample. The presence of intragranular pores would pin down the magnetic moment in grains, thus reducing the M_s and also the H_c. The decrease in H_c as pH increases can be attributed to the decrement of magnetocrystalline anisotropy with anisotropic Fe^{2+} ions located in a 2A site, and the enlargement of the grain size is evident in FESEM micrographs (**Figure 5**). The M_s and M_r are also observed to decrease as pH increases. The decrement of magnetic parameters as pH increases could be due to the existence of large amount of diamagnetic phases as the amount of ammonia NH_3 increases. It seems that the main roles of the diamagnetic NH_3 are to isolate Sr-ferrite nanoparticles from each other, thus reducing exchange interaction between them, and are known to have a detrimental effect on M_s and M_r.

4. Conclusions

Single-phase $SrFe_2O_{19}$ ferrite nanoparticles were successfully synthesized by sol-gel citrate-nitrate method. From the discussion presented earlier, the influence of pH variation on the $SrFe_2O_{19}$ ferrite nanoparticles on the structural, microstructural, and magnetic properties was discussed. An increment amount of ammonia has changed the purity, average grain size, density, and its porosity, which affected the magnetic properties of the samples. Those characteristics reveal an understanding on how important effects of pH study (linear effect of pH and acidic-alkaline effect) underlining on $SrFe_{12}O_{19}$ nanoparticles, as most researchers neglect it.

Acknowledgements

The authors would like to thank the Ministry of Education Malaysia for providing funds; MyBrain15, Research University Grants Vot No. 9541600 and 5524942; and the Department of Physics, Faculty of Science and the Materials, Synthesis and Characterization Laboratories (MSCL) ITMA, UPM, for the measurement facilities.

Conflict of interest

The authors declare that they have no competing interest.

Author details

Muhammad Syazwan Mustaffa[2], Rabaah Syahidah Azis[1,2]* and Sakinah Sulaiman[1]

*Address all correspondence to: rabaah@upm.edu.my

1 Materials Synthesis and Characterization Laboratory, Institute of Advanced Technology, Universiti Putra Malaysia, Serdang, Selangor, Malaysia

2 Department of Physics, Faculty of Science, Universiti Putra Malaysia, Serdang, Selangor, Malaysia

References

[1] Masoudpanah SM, Ebrahimi SAS, Ong CK. Preparation of strontium hexaferrite film by pulsed laser deposition with *in situ* heating and post annealing. Journal of Magnetism and Magnetic Materials. 2012;**324**(18):2894-2898. DOI: 10.1016/j.jmmm.2012.04.034

[2] Masoudpanah SM, Ebrahimi SAS. Synthesis and characterization of nanostructured strontium hexaferrite thin films by the sol-gel method. Journal of Magnetism and Magnetic Materials. 2012;**324**(14):2239-2244. DOI: 10.1016/j.jmmm.2012.02.109

[3] Perdamean S, Muljadi M, Siregar RT, Tomi BW. Ferrite-based material as a permanent magnet for components of electrical generators. Advances in Natural Sciences: Nanoscience and Nanotechnology. 2011;**2**:045016. DOI: 10.1088/2043-6262/2/4/045016

[4] Topal U, Bakan HI. Permanently magnetic $BaFe_{12}O_{19}$ foams: Synthesis and characterization. Materials Chemistry and Physics. 2010;**123**(1):121-124. DOI: 10.1016/j.matchemphys.2010.03.070

[5] Ashiq MN, Iqbal MJ, Najam-Ul-Haq M, Hernandez Gomez P, Qureshi AM. Synthesis, magnetic and dielectric properties of ErNi doped Sr-hexaferrite nanomaterials for applications in high density recording media and microwave devices. Journal of Magnetism and Magnetic Materials. 2012;**324**(1):15-19. DOI: 10.1016/j.jmmm.2011.07.016

[6] Hibst H, Schwab E. Magnetic recording materials. Materials Science and Technology. Federal Republic of Germany: WILEY-VCH Verlag GmbH & CoKGaA; 2006. DOI: 10.1002/9783527603978.mst0041

[7] Minh T, Dang H, Trinh VD, Bui DH. Sol-gel hydrothermal synthesis of strontium hexaferrite nanoparticles and the relation between their crystal structure and high coercivity properties. Advances in Natural Science and Nanotechnology. 2012;**3**(2):1-7. DOI: 10.1088/2043-6262/3/2/025015

[8] Singhal S, Namgyal T, Singh J, Chandra K, Bansal S. A comparative study on the magnetic properties of $MFe_{12}O_{19}$ and $MAlFe_{11}O_{19}$ (M = Sr, Ba and Pb) hexaferrites with different

morphologies. Ceramics International. 2011;**37**(6):1833-1837. DOI: 10.1016/j.ceramint.2011.02.001

[9] Pullar RC. Hexagonal ferrites: A review of the synthesis, properties and applications of hexaferrite ceramics. Progress in Materials Science. 2012;**57**(7):1191-1334. DOI: 10.1016/j.pmatsci.2012.04.001

[10] Liu JP, Fullerton E, Gutfleisch O, Sellmyer DJ. Nanoscale Magnetic Materials and Applications. London, New York: Springer Dordrecht Heidelberg; 2010. DOI: 10.1007/978-0-387-85600-1

[11] Sharma P, Rocha RA, Medeiros SN, Hallouche B, Paesano A. Structural and magnetic studies on mechanosynthesized $BaFe_{12-x}Mn_xO_{19}$. Journal of Magnetism and Magnetic Materials. 2007;**316**:29-33. DOI: 10.1016/j.jmmm.2007.03.207

[12] Sánchez-DeJesús F, Bolarín-Miró AM, Cortés-Escobedo CA, Valenzuela R, Ammar S. Mechanosynthesis, crystal structure and magnetic characterization of M-type $SrFe_{12}O_{19}$. Ceramics International. 2014;**40**:4033-4038

[13] Luo J. Structural and magnetic properties of Nd-doped strontium ferrite nanoparticles. Materials Letters. 2012;**8**:162-164

[14] Faloh-Gandarilla JC, Dıaz-Castanon S, Watts BE. Magnetization reversal and interactions in $SrFe_{12}O_{19}$. Physica Status Solidi B. 2016;**254**(4):1-7. https://doi.org/10.1002/pssb.201600393

[15] Ashiq MN, Iqbal MJ, Gul IH. Structural, magnetic and dielectric properties of Zr-Cd substituted strontium hexaferrite ($SrFe_{12}O_{19}$) nanoparticles. Journal of Alloys and Compounds. 2009;**487**:341-345

[16] Kostishyn VG, Panina LV, Kozhitov LV, Timofeev AV, Kovalev AN. Synthesis and multiferroic properties of M-type $SrFe_{12}O_{19}$ hexaferrite ceramics. Journal of Alloys and Compounds. 2015;**645**:297-300

[17] Wong YC, Wang J, Teh GB. Structural and magnetic studies of $SrFe_{12}O_{19}$ by sol-gel method. Procedia Engineering. 2014;**76**:45-52. DOI: 10.1016/j.proeng.2013.09.246

[18] Zhang Z, Liu X, Wang X, Wu Y, Li R. Effect of Nd-Co substitution on magnetic and microwave absorption properties of $SrFe_{12}O_{19}$ hexaferrites. Journal of Alloys and Compounds. 2012;**525**:114-119

[19] Afghahi SSS, Peymanfar R, Javanshir S, Atassi Y, Jafarian M. Synthesis, characterization and microwave characteristics of ternary nanocomposite of MWCNTs/doped Sr-hexaferrite/PANI. Journal of Magnetism and Magnetic Materials. 2016;**423**:152-157

[20] Xia A, Zuo C, Chen L, Jin C, Lv Y. Hexagonal $SrFe_{12}O_{19}$ ferrites: Hydrothermal synthesis and their sintering properties. Journal of Magnetism and Magnetic Materials. 2013;**332**(4):186-191

[21] Mirkazemi SM, Alamolhoda S, Ghiami Z. Erratum to: Microstructure and magnetic properties of $SrFe_{12}O_{19}$ nano-sized powders prepared by sol-gel auto-combustion method with

CTAB surfactant. Journal of Superconductivity and Novel Magnetism. 2015;**28**(5):1551-1558. DOI: 10.1007/s10948-014-2872-x

[22] Liu Y, Drew MG, Liu Y. Preparation and magnetic properties of barium ferrites substituted with manganese, cobalt, and tin. Journal of Magnetism and Magnetic Materials. 2007;**323**:945-953. DOI: 10.1016/j.jmmm.2010.11.075

[23] Nazlan R, Hashim M, Ibrahim IR, Mohd Idris F, Wan Ab Rahman WN, Abdullah NH, et al. Influence of indium substitution and microstructure changes on the magnetic properties evolution of $Y_3Fe_{5-x}In_xO_{12}$ (x = 0.0 − 0.4). Journal of Materials Science: Materials in Electronics. 2015;**26**(6):3596-3609. DOI: 10.1007/s10854-015-2874-x

[24] Fatemeh B, Masoud R, Mehdi MP. Synthesis of $SrFe_{12}O_{19}/SiO_2/TiO_2$ composites with core/shell/shell nano-structure and evaluation of their photo-catalytic efficiency for degradation of methylene blue. Journal of Materials Science: Materials in Electronics. 2018;**29**(3):1877-1887

[25] Mangai KA, Sureshkumar P, Priya M, Rathnakumari M. Structural and magnetic properties of strontium hexa-ferrites for permenant magnets. International Journal of Scientific & Engineering Research. 2014;**5**(3):65-69

[26] Masoudpanah SM, Seyyed Ebrahimi SA. Effect of pH value on the structural and magnetic properties of nanocrystalline strontium hexaferrite thin films. Journal of Magnetism and Magnetic Materials. 2011;**323**(21):2643-2647. DOI: 10.1016/j.jmmm.2011.05.055

[27] Dang TMH, Trinh VD, Bui DH, Phan MH, Huynh DC. Sol-gel hydrothermal synthesis of strontium hexaferrite nanoparticles and the relation between their crystal structure and high coercivity properties. Advances in Natural Sciences: Nanoscience and Nanotechnology. 2012;**3**(2):25015

[28] Wang Y, Li Q, Zhang C, Li B. Effect of Fe/Sr mole ratios on the formation and magnetic properties of $SrFe_{12}O_{19}$ microtubules prepared by sol-gel method. Journal of Magnetism and Magnetic Materials. 2009;**321**(19):3368-3372. DOI: 10.1016/j.jmmm.2009.05.066

[29] Reza G, Ghasemi A, Saidi A. Enhanced magnetic properties of substituted Sr-hexaferrite nanoparticles synthesized by co-precipitation method. Ceramics International. 2014;**40**(3):4945-4952. DOI: 10.1016/j.ceramint.2013.10.096

[30] Malhotra S, Chitkara M, Sandhu IS. Microwave absorption study of nano synthesized strontium ferrite particles in X band. International Journal of Signal Processing, Image Processing and Pattern Recognition. 2015;**8**(10):115-120

[31] Maswadeh Y, Mahmood SH, Awadallah A, Aloqaily AN. Synthesis and structural characterization of nonstoichiometric barium hexaferrite materials with Fe:Ba ratio of 11.5–16.16. IOP Conference Series: Materials Science and Engineering. 2015;**92**:012019. DOI: 10.1088/1757-899X/92/1/012019

[32] Pereira FMM, Junior CAR, Santos MRP, Sohn RSTM, Freire FNA, ki JM, et al. Structural and dielectric spectroscopy studies of the M-type barium strontium hexaferrite alloys

($Ba_xSr_{1-x}Fe_{12}O_{19}$). Journal of Materials Science: Materials in Electronics. 2008;**19**(7):627-638. DOI: 10.1007/s10854-007-9411-5

[33] Song F, Shen X, Xiang J, Song H. Formation and magnetic properties of M-Sr ferrite hollow fibers via organic gel-precursor transformation process. Materials Chemistry and Physics. 2010;**120**(1):213-216. DOI: 10.1016/j.matchemphys.2009.10.048

[34] Sivakumar M, Gedanken A, Zhong W, Du YW, Bhattacharya D. Nanophase formation of strontium hexaferrite fine powder by the sonochemical method using $Fe(CO)_5$. Journal of Magnetism and Magnetic Materials. 2004;**268**:95-104. DOI: 10.1016/S0304-8853(03)00479-7

[35] Nazlan R, Hashim M, Abdullah NH, Ibrahim IR, Ismail I. Influence of milling time on the crystallization, morphology and magnetic properties of polycrystalline yttrium iron garnet. Advanced Materials Research. 2012;**501**:324-328. DOI: 10.4028/www.scientific.net/AMR.501.324

[36] Yang FJ, Min JJ, Kang ZW, Tu SY, Chen HB, Liu DG, et al. The influence of pH value and composition on the microstructure, magnetic properties of Co-Fe-Al Heusler nanoparticles. Chemical Physics Letters. 2017;**670**(3):1-4. DOI: 10.1016/j.cplett.2016.12.062

[37] Tian L, Xu J, Xiao S. The influence of pH and bath composition on the properties of Ni-Co coatings synthesized by electrodeposition. Vacuum. 2011;**86**(1):27-33. DOI: 10.1016/j.vacuum.2011.03.027

[38] Hesani M, Yazdani a, Abedi Ravan B, Ghazanfari M. The effect of particle size on the characteristics of FeCo nanoparticles. Solid State Communications. 2010;**150**(13–14):594-597. DOI: 10.1016/j.ssc.2009.12.043

Chapter 5

Synthesis of Conductive Sol-Gel ZnO Films and Development of ZnO Printed Electronics

David Winarski and Farida Selim

Additional information is available at the end of the chapter

http://dx.doi.org/10.5772/intechopen.82041

Abstract

ZnO thin films are synthesized and studied to understand the functionality of solution-processed semiconductor devices. A simple sol-gel technique is used to fabricate transparent conductive oxides (TCOs) and ultraviolet (UV) photodetectors from ZnO precursors via spin coating, inkjet printing (IJP), and aerosol jet printing (AJP). A variety of flexible and transparent substrates was selected based on the deposition and sintering conditions and the device application. Doping of ZnO films with Al^{3+}, In^{3+}, and Ga^{3+} was introduced in precursor solutions before deposition processes. Post-deposition processing was carried out in air, H_2, and Zn environments to optimize thin film properties. Optical, structural, and electronic data analyses reveal the significant effects that deposition method, substrates, dopants, and processing conditions have on the optical transmission, crystallinity, grain size, and electrical conductivity.

Keywords: flexible electronics, zinc oxide, UV photodetectors, aerosol jet printing, inkjet printing, vacancy passivation, positron annihilation spectroscopy

1. Introduction

Printing electrically functional inks has emerged as an important research topic to drive device technologies into the future. It has some advantages compared to conventional fabrication techniques in terms of low cost and applicability for flexible devices. This is promising for wearable, implantable, patch-like, and textile-integrated electronics, advancing the device field. With the right ink and substrate, it would be possible to achieve lightweight, flexible, transparent devices with a good electrical performance, which will revolutionize our daily lives.

Until recently, organic inks have monopolized printing technology because of their printability, flexibility, and electronic functionality. However, developing printable inorganic inks would allow for higher performance—like conventional devices—at a much lower cost than conventional fabrications, such as atomic layer deposition (ALD), pulsed laser deposition (PLD), chemical vapor deposition (CVD), physical vapor deposition (PVD), molecular beam epitaxy (MBE), and sputtering. In addition, printing techniques allow for deposition at low temperatures and at specified locations, a controllability of both parameters that no other deposition techniques can claim. These advantages make printing much easier and more compatible with flexible substrates, as some polymers cannot withstand the high processing temperatures of some deposition techniques or the harsh chemicals and UV radiation in the photolithography necessary to construct a functional device.

Some facile printing technologies are inkjet (IJP) and aerosol jet printing(AJP). Inkjet printing (IJP) is very quick and a simple drop-on-demand or continuous stream technique with micrometer precision [1, 2]. Aerosol jet printing (AJP) is favored as a clean and precise technique, using a continuous stream to print features down to 10 μm [3]. Nonetheless, both techniques are new and exciting ways to fabricate electronic devices. Both AJP and IJP techniques are currently used to print a wide variety of organic and inorganic inks for use as flexible photodetectors, transistors, and other circuit board components [4–10]. ZnO is an exciting material for electronics due to its direct wide bandgap (3.2 eV at 298 K), strong UV absorption, and electrical tunability. Many researchers have successfully fabricated ZnO devices at low cost and relatively low temperature by way of printing sol-gel precursor and other nanoparticle-based inks [11–20]. In addition, the authors of this chapter have successfully fabricated ZnO transparent conductive oxides (TCOs) using a simple sol-gel, spin coating technique [21]. Research efforts utilizing sol-gel-derived ZnO thin films for device applications have greatly increased, recently [22–24]. Through these methods, the resultant ZnO material properties can be tuned by introducing group III metal ions during the precursor sol synthesis. In this chapter, we present the use of a sol-gel technique to develop ZnO printed electronics.

To understand the applications of ZnO sol-gel precursors, we employ spin coating, IJP, and AJP deposition with thermal and photonic sintering to synthesize and tune ZnO thin films. Extrinsic shallow donors, such as Al^{3+}, In^{3+}, and Ga^{3+}, have similar ionic radius to that of Zn^{2+} and thus can easily replace it with little effect on the lattice structure [20, 25–29]. These donors are introduced during the precursor solutions' synthesis and various atmospheric post-processing heat treatments are applied to introduce, eliminate, or passivate intrinsic defects that greatly alter the electrical conductivity of the films. The ZnO thin film properties are studied by scanning electron microscopy (SEM), X-ray diffraction (XRD), ultraviolet-visible range (UV-VIS) absorbance, Van der Pauw and Hall effect measurements, and positron annihilation spectroscopy (PAS).

2. Sol-gel synthesis

To make a ZnO sol-gel precursor, we dissolve zinc acetate (99.99%) in 2-methoxyethanol (99.8%)—using ethanolamine (99%) as a stabilizer—to obtain a 0.75 M solution, with zinc acetate and ethanolamine at a 1:1 molar ratio. To dope ZnO thin films, aluminum(III) nitrate nonahydrate (99.997%), gallium(III) nitrate hydrate (99.9998%), and indium(III) acetate

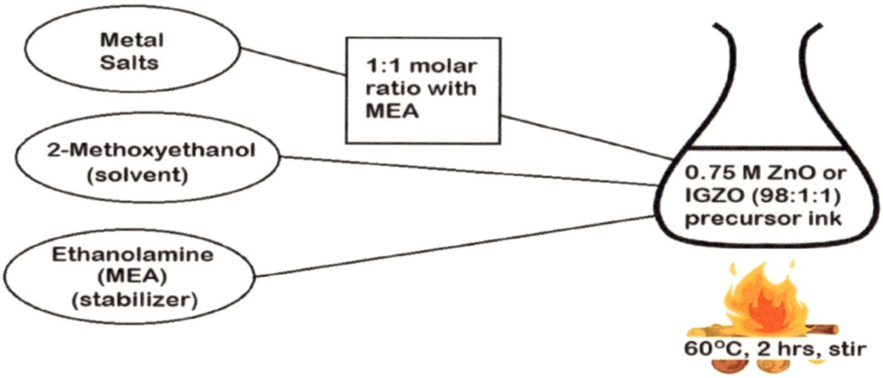

Figure 1. Synthesis procedure for ZnO, AZO, GZO, and IGZO precursor solutions.

hydrate (99.99%) metal salts were implemented to replace some zinc acetate in the mixture to obtain a doping level of 1% in solution while keeping the 1:1 molar ratio with ethanolamine and the molarity at 0.75 M. Undoped ZnO, aluminum-doped ZnO (AZO), gallium-doped ZnO (GZO), and indium-gallium-codoped ZnO (IGZO) precursor solutions were prepared in an open-air environment, then covered with plastic paraffin film, heated to 60°C, and magnetically stirred for 2 h to obtain a transparent homogenous solution and then left to cool before deposition (**Figure 1**).

3. Substrate preparation: piranha etching and atmospheric plasma treatment

Quartz, cyclic olefin copolymer (TOPAS), polyethylene terephthalate (PET), and polyimide (Kapton) substrates were selected based on their transparency and/or flexibility. Before ZnO deposition, substrates were cleaned and etched to improve substrate/solution compatibility. Quartz substrates were treated in piranha baths to clean residual contaminants from the substrate and induce a surface charge. The substrate was placed in a 3:1 H_2SO_4:H_2O_2 bath at 80°C for 15 min, rinsed with deionized water, placed in a 3:1 NH_4OH:H_2O_2 bath at 80°C for 15 min, rinsed with deionized water again, and placed in an oven at about 100°C to dry. Before IJP and AJP, quartz, TOPAS, PET, and Kapton substrates were prepared by swabbing with acetone and isopropanol, drying with a nitrogen gun, and applying atmospheric plasma treatment from a corona discharge wand with the transformer set at 200 W at a standoff distance of 5 mm from the ground electrode.

4. Thin film synthesis

Once the sol-gel precursors and substrates have been prepared, ZnO thin films were deposited using spin coating, IJP, and AJP. **Table 1** summarizes ZnO sol-gel precursors, substrates, deposition methods, sintering conditions, and post-processing environments and thicknesses.

4.1. Spin coating

A Laurell Technologies Corporation spin coater was used to spin a quartz substrate at 500 rpm. Then, 40–50 drops of precursor solution were dispensed, before the substrate/solution was accelerated to 3000 rpm and left spinning for 30 s to obtain a gel-like thin layer. Next, the gel film was placed in an oven to dry at 150°C for 10 min. The spin coating and drying processes were repeated to obtain the desired number of layers (10–16 layers total). Finally, the films were annealed in ambient air at 400°C for 60 min, to obtain a ZnO wurtzite structure. ZnO, AZO, and GZO films were fabricated using this spin coating technique. To tune the electronic properties, several samples were further annealed in the following flowing gas conditions: (1) forming gas of 95% N_2 and 5% H_2 at 400°C for 60 min, (2) H_2 flow at 400°C for 60 min, and (3) Zn-rich environment in Ar at 400°C for 60 min. The Zn-rich environment was created with Zn powder (99.999%) and Zn foil (99.994%, 0.1 mm thick). Thin films and Zn powder were wrapped tightly in Zn foil, while an Ar gas flow was used to prevent oxidation of the ZnO.

4.2. Inkjet printing

A Dimatix inkjet printer printed 7 mm × 7 mm squares of ZnO and IGZO sol-gel precursors onto various substrates for a total of 12 layers. The jet and platen temperatures were set to

Sample	Substrate	Deposition method	Sintering conditions	Thickness (nm)
GZO	Quartz	Spin coating	400°C, 60 min, air	800
GZO	Quartz	Spin coating	400°C, 60 min, air	808
AZO	Quartz	Spin coating	400°C, 60 min, air	515
ZnO^{0}	Quartz	Spin coating	400°C, 60 min, air	600
ZnO^{0}	Quartz	Spin coating	400°C, 60 min, air	173
ZnO^{1}	TOPAS	Inkjet printing	170°C, 60 min, air	~600
ZnO^{2}	Kapton	Inkjet printing	300°C, 20 min, air	~600
ZnO^{3}	TOPAS	Inkjet printing	Xenon, 180 bursts, N_2	~600
ZnO^{3}	Kapton	Inkjet printing	Xenon, 180 bursts, N_2	~600
ZnO^{3}	PET	Inkjet printing	Xenon, 180 bursts, N_2	~600
ZnO^{4}	TOPAS	Inkjet printing	150°C, 30 min, air	~600
ZnO^{4}	Kapton	Inkjet printing	150°C, 30 min, air	~600
ZnO^{5}	Kapton	Inkjet printing	400°C, 60 min, air	~600
ZnO^{6}	Kapton	Aerosol jet printing	200°C, 60 min, air	~400
ZnO^{7}	Kapton	Aerosol jet printing	300°C, 60 min, air	~400
ZnO^{8}	Kapton	Aerosol jet printing	400°C, 60 min, air	~400
IGZO^{1}	Quartz	Inkjet printing	400°C, 60 min, air	~600
IGZO^{2}	Kapton	Inkjet printing	400°C, 60 min, air	~600
IGZO^{3}	Kapton	Aerosol jet printing	400°C, 60 min, air	~400

Table 1. A summary of ZnO sol-gel precursors, substrates, deposition methods, sintering conditions, and post-processing environments and thicknesses. Thicknesses were evaluated by SEM, ellipsometry, and profilometry techniques [20, 21].

39°C, and the droplet overlap was set to at least 50%. The droplet size was between 50 and 100 μm, depending on the substrate. The resultant gels were dried at 150°C for 10–30 min (until visibly dry) to remove any residual solvent. Here, processing techniques were limited by the thermal expansion coefficient of the substrates. Post-print sintering was carried out using thermal and photonic sintering methods. Thermal sintering took place in ambient atmosphere on a hot plate at temperatures between 170 and 400°C for 20–60 min. Photonic sintering was carried out in a N_2-rich atmosphere using a xenon arc lamp placed 4.445 cm above the substrate platen, set at 2 kV with 6-ms pulse width for a total of 180 bursts.

4.3. Aerosol jet printing

An Optomec aerosol jet printer printed similar 7 mm × 7 mm squares of ZnO sol-gel precursors onto Kapton substrates for a total of six layers. A 200-μm nozzle was used at a speed of 3 mm/s. The line width of the aerosol spray was about 75 μm, so a 50-μm serpentine pattern was selected to achieve ~33% overlap. The resultant gels were dried at 90°C for 30 min (until visibly dry), then subject to thermal sintering in ambient atmosphere on a hot plate at 200°, 300°, and 400°C for 60 min.

5. Scanning electron microscopy

Spin-coated AZO thin films that were post-processed in H_2 and Zn were imaged by SEM. Low-magnification surface images show worm-like structures (**Figure 2**), while a higher

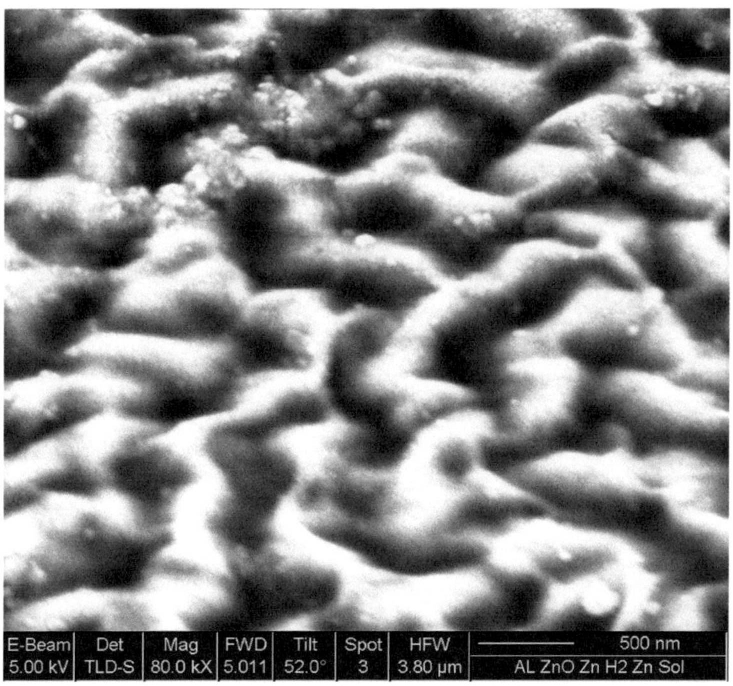

Figure 2. Low-magnification SEM surface image for AZO film deposited by spin coating and post-processed in H_2 and Zn environments, consecutively [21].

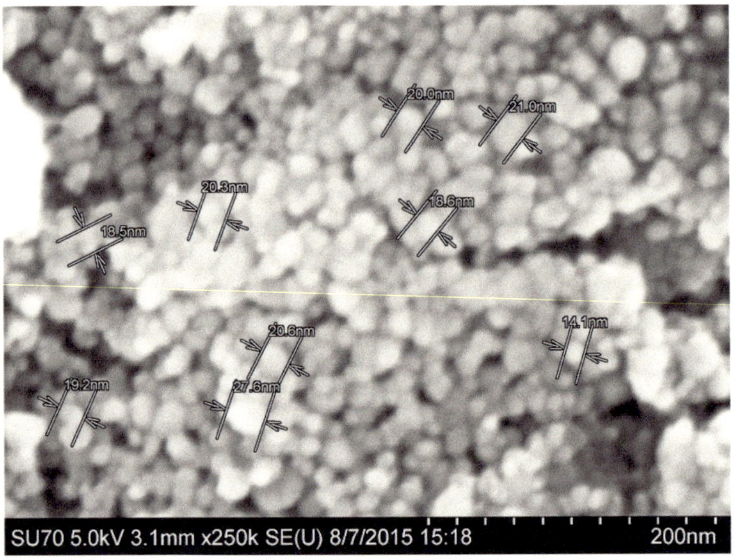

Figure 3. High-magnification SEM surface image for AZO film deposited by spin coating and post-processed in H_2 and Zn environments, consecutively [21].

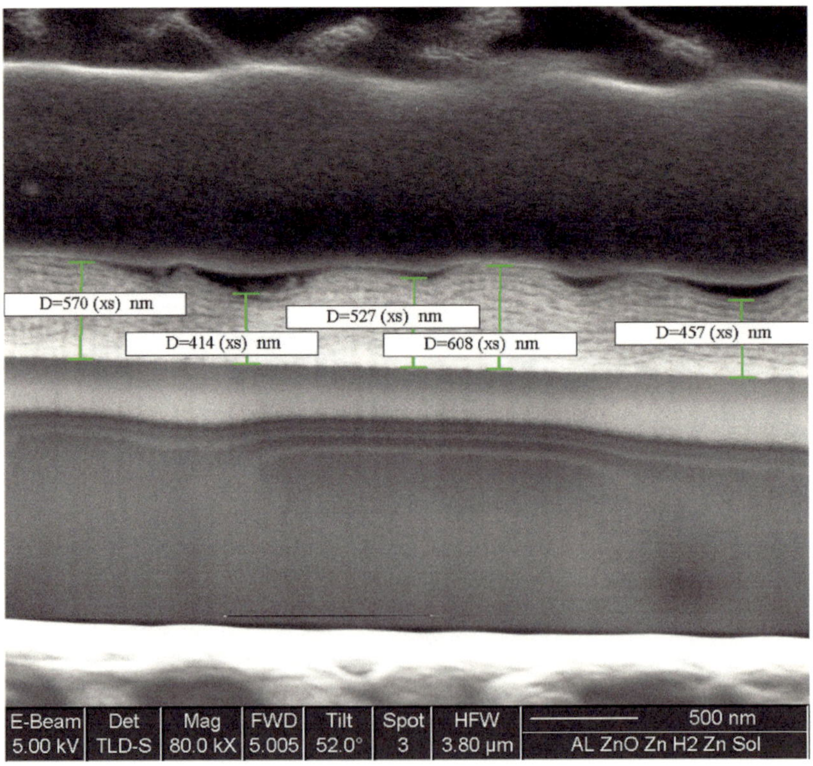

Figure 4. High-magnification cross-sectional SEM image of AZO films deposited by spin coating and post-processed in H_2 and Zn environments, consecutively [21].

magnification shows round particles with an average particle size of 20 nm (**Figure 3**). Platinum was then deposited on the film surface by a focused ion beam and a trench was milled through the sample to obtain a high-resolution cross-sectional image (**Figure 4**). It can be seen that the film is deposited as distinct individual layers (each layer is ~40 nm thick). The images also reveal non-uniform thickness, with ~25% variation across the film and indicate that it is difficult to obtain uniform thickness using sol-gel methods. These images represent the first high-resolution cross-sectional images for sol-gel films. They illustrate that the distinct individual layers and the non-uniformity in thickness are inherent of the spin coating method, but they may be reduced by further annealing. This non-uniform layering leads to interference in UV-VIS transmission spectra, a well-known feature in sol-gel films.

6. X-ray diffraction

Film crystallinity was studied using a Rigaku X-ray diffractometer to determine the ZnO crystal phase (*hkl* values) and average grain size. XRD patterns for spin-coated AZO films indicate polycrystalline thin films, with peaks corresponding to the (100), (002), and (101) planes (**Figure 5**). These diffraction patterns match the ZnO hexagonal wurtzite structure, without any secondary phase impurities in the films. Furthermore, we can see a change in polycrystallinity based on post-processing in H_2 and Zn environments. When processed in H_2, films become less polycrystalline, while processing in Zn leads to the opposite effect.

XRD patterns for IJP ZnO films (**Figure 6**) reveal that ZnO phases form at temperatures as low as 150°C, with more peaks appearing at increased sintering temperature. Although there is a lack of ZnO phase formation for ZnO[4] on Kapton and an impurity phase for ZnO[3] on TOPAS,

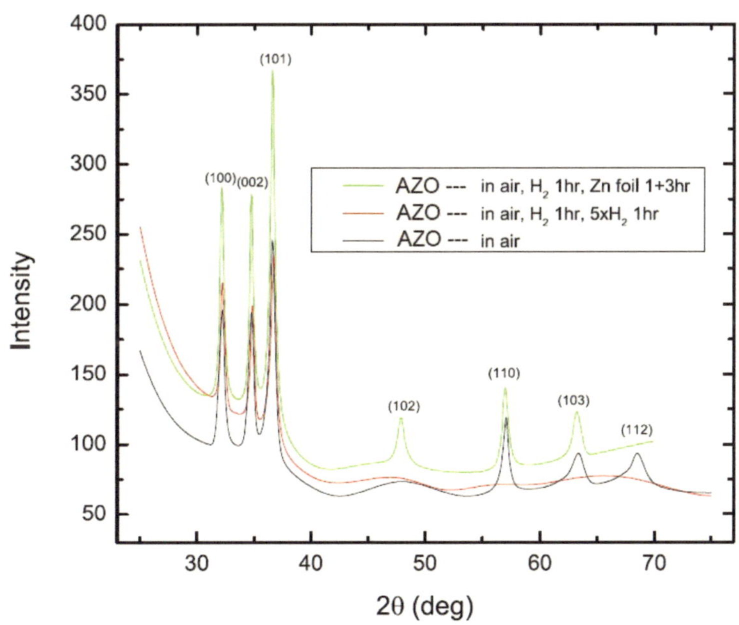

Figure 5. XRD measurements for AZO films annealed in various atmospheres [21].

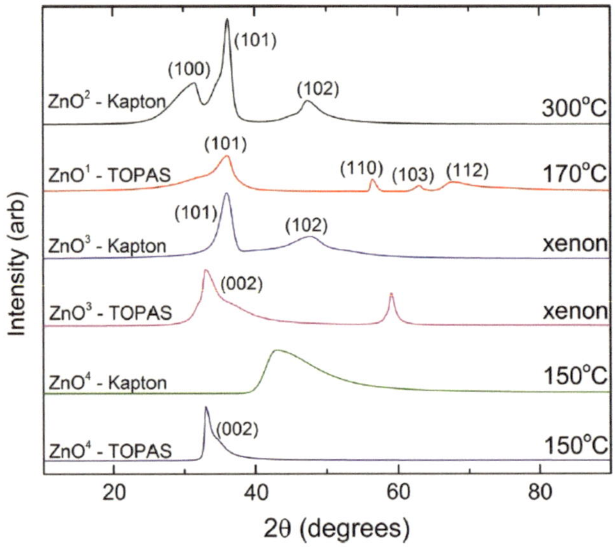

Figure 6. XRD spectra for IJP films on Kapton and TOPAS substrates sintered by hot plate and xenon arc lamp [20].

the results demonstrate that ZnO thin films can be successfully fabricated by inkjet printing and thermal and photonic sintering processes.

XRD patterns for AJP ZnO reveal amorphous nature at 200°C and an increasing polycrystallinity with the sintering temperature. In addition, increasing the sintering temperature increases the average grain size, which is consistent with previous reports [30, 31]. XRD patterns for AJP IGZO XRD show a ~50% increase in grain size due to the low doping concentration of In^{3+} and Ga^{3+}, with a minimal effect on the polycrystalline structure. **Table 2** presents the average grain sizes of the aforementioned thin films. The grain size D was calculated for each 2θ peak using the Scherrer equation:

$$D = 0.9\lambda/\beta \cos\theta, \tag{1}$$

where $\lambda = 1.54$ Å is the X-ray wavelength, β is the full width at half maximum (FWHM) of the corresponding peak, and θ is the Bragg angle.

Sample	Sintering conditions	Average grain size (Å)	Estimated standard deviation
ZnO^6	200°C, 60 min, air	32.33	6.04
ZnO^7	300°C, 60 min, air	239.17	36.25
ZnO^8	400°C, 60 min, air	500.32	119.91
$IGZO^3$	400°C, 60 min, air	736.84	27.85

Table 2. Average grain size—calculated by the Scherrer equation—and standard error values corresponding to ZnO and IGZO films in **Figure 7** [20].

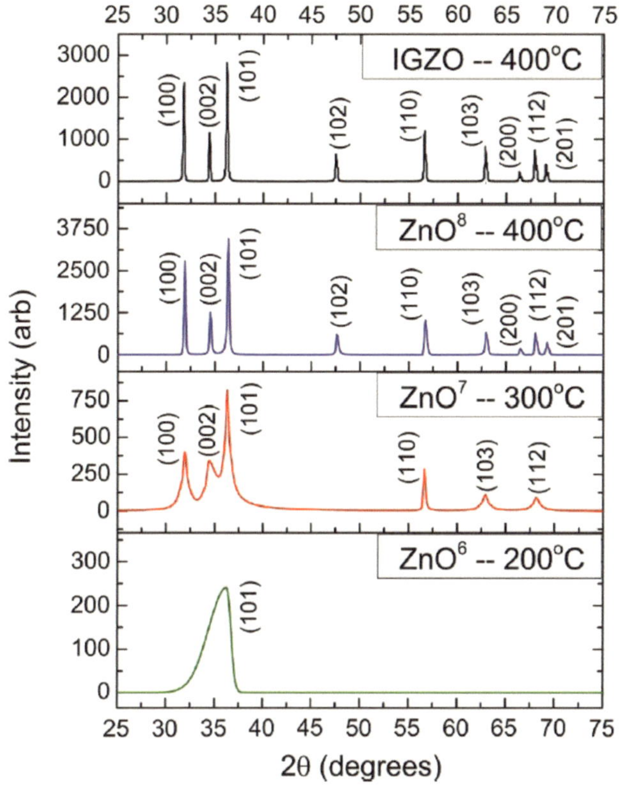

Figure 7. XRD spectra for AJP ZnO films on Kapton sintered at 200°C, 300°C, and 400°C and AJP IGZO films on Kapton sintered at 400°C [20].

ZnO, AZO, GZO, and IGZO sol-gel precursors are viable options to achieve a ZnO wurtzite structure at low sintering temperatures. Films are generally inhomogeneous in thickness and amorphous or polycrystalline in nature, with grain size and polycrystallinity increasing with the sintering temperature. A low doping concentration does not inhibit ZnO wurtzite formation, but the incorporation of In^{3+} and Ga^{3+} dopants effectively increases the average grain size. Furthermore, post-processing in H_2 and Zn environments can change the polycrystallinity of the films.

7. Ultraviolet-visible range spectroscopy

A dual-beam Perkin Elmer UV-VIS spectrometer was used to record the transmission and absorbance spectra of spin-coated and printed ZnO films. A blank substrate was placed in line with that reference beam, while the sample spectra were recorded.

Transmission measurements for spin-coated ZnO^0 and AZO show the band edge near 380 nm with a high visible range transparency (**Figure 8**). The individual layering, as observed from

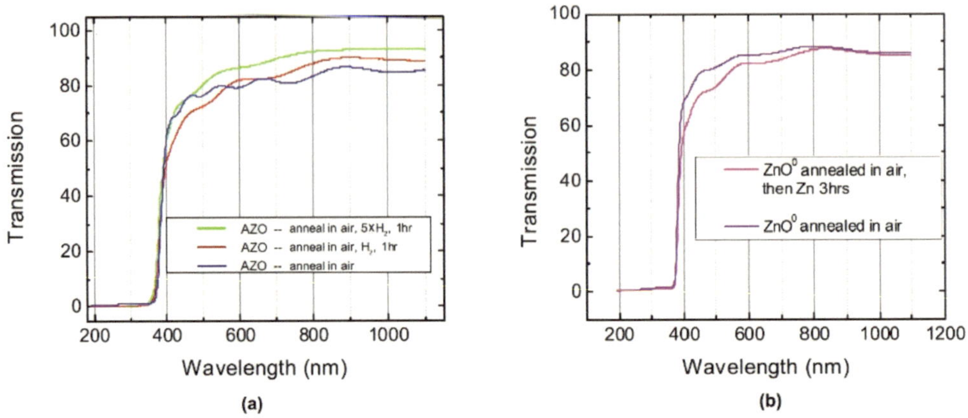

Figure 8. UV-VIS transmission measurements for: (a) AZO films before and after hydrogen treatment at different pressures and (b) ZnO[0] before and after Zn treatment [21].

SEMs, leads to interference effects in the spectra, which can be reduced by a greater H_2 concentration during annealing and lead to improved transparency (**Figure 8a**). This can be explained by a decrease in polycrystallinity observed in XRD analysis. However, the opposite effect occurs after annealing in a Zn environment (**Figure 8b**), which is due to the increase in polycrystallinity.

Printed ZnO and IGZO films also show a band edge near 380 nm from optical absorbance measurements (**Figure 9a**). The bandgap is near 3.2 eV for all printed films, as calculated by the Tauc method (**Figure 9b**). AJP IGZO films are more transparent than IJP films due to better overall print quality. **Figure 10** compares the IJP and AJP techniques, as seen by the naked eye. It is clear that AJP films are more transparent because of less light scattering from surface roughness and striations in the IJP films.

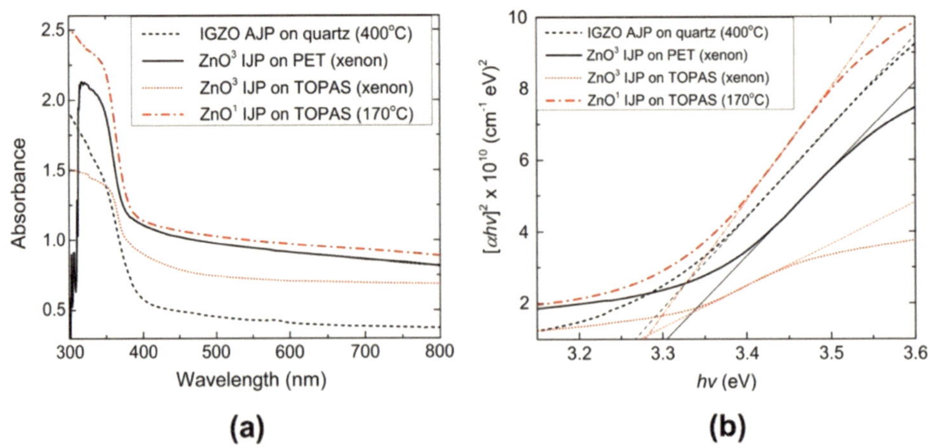

Figure 9. (a) UV-VIS absorbance spectra for IJP and AJP ZnO and IGZO films sintered by different methods, exhibiting a band edge near 380 nm and (b) Tauc plots of direct-bandgap transitions for each spectrum with linear fits extrapolated to $(\alpha h\nu)^2 = 0$ for bandgap determination [20].

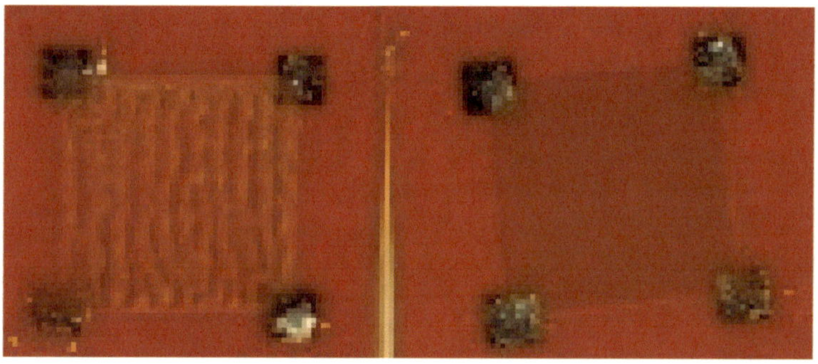

Figure 10. Comparison of ZnO thin films printed by IJP (*left, ZnO^2*) and AJP (*right, ZnO^7*) sintered at 300°C. The printed silver pads and indium contacts were applied to the corners for Hall effect measurements. Image taken by a Samsung Galaxy S6 [20].

These results established that the sol-gel precursor method can produce films with good visible range transparency in spin coating and printing techniques. Interference in the visible range absorbance can be reduced by the post-processing conditions. Here, AJP yields a better print quality than IJP and offers similar visible range transparency to spin coating.

8. Van der Pauw and Hall effect measurements

At 300 K, the resistivity was obtained via van der Pauw measurements using an MMR Hall effect system. All spin-coated and printed ZnO, AZO, GZO, and IGZO films show high electrical resistivity after the initial sintering. All spin-coated films were too resistive to initially measure, and the printed films measured resistivity $>10^4$ Ω cm. However, post-processing of spin-coated GZO, AZO, and ZnO^0 in H_2 and Zn environments induced a large conductivity. Van der Pauw and Hall effect measurements for ZnO, AZO, GZO, and IGZO films are summarized in **Table 3**.

AZO thin films annealed in both H_2 and Zn offer the lowest electrical resistivity (1.71×10^{-2} Ω cm) and the highest carrier concentration (3.01×10^{21}). We emphasize that the electrical conductivity results only after the post-processing steps. The large decrease in resistivity is attributed to the passivation of defect states, which will be discussed further in the PAS section of this chapter. The low resistivity coupled with the high visible range transparency offers solution-processed ZnO as a viable option of TCO in printed electronics.

In^{3+} and Ga^{3+} dopants were also investigated through electrical measurements by comparing printed $IGZO^2$ and ZnO^5 thin films. Both use IJP deposition on Kapton substrates and have a thickness of 600 nm. Unsurprisingly, $IGZO^2$ has a lower resistivity (3.06×10^4 Ω cm) than ZnO^5 (4.59×10^5 Ω cm). It is well understood that In^{3+} and Ga^{3+} increase conductivity in ZnO, an effect studied in other In- and Ga-doped sol-gel ZnO [32, 33].

The lowest resistivity for AJP ZnO is in the thin film that was annealed at 300°C. The overall resistivity of ZnO depends on its structural properties, which are affected by the oxygen concentration in the film. At higher sintering temperatures, resistivity increases with annealing

Sample	Post-processing conditions	Resistivity (Ω cm)	Mobility ($cm^2 V^{-1} s^{-1}$)	Carrier concentration (cm^{-3})
GZO	400°C, 60 min, H_2	1.03×10^1	<1	1.43×10^{19}
GZO	400°C, 60 min, H_2 & 400°C, 60 min, Zn	1.97×10^{-1}	<1	1.44×10^{20}
AZO	400°C, 60 min, H_2 & 400°C, 240 min, Zn	1.71×10^{-2}	<1	3.01×10^{21}
ZnO[0]	400°C, 60 min, H_2/N_2 & 400°C, 180 min, Zn	1.83×10^{-1}	2.94×10^1	1.16×10^{18}
ZnO[0]	400°C, 60 min, Zn	1.08×10^2	<1	1.86×10^{17}
ZnO[5]	—	4.59×10^5	—	—
IGZO[2]	—	3.06×10^4	—	—
ZnO[6]	—	1.02×10^5	—	—
ZnO[7]	—	8.36×10^4	—	—
ZnO[8]	—	2.25×10^5	—	—

Table 3. Van der Pauw and Hall effect measurements for ZnO, AZO, GZO, and IGZO thin films grown from ZnO sol-gel precursors, listing the film type, post-processing conditions, resistivity, mobility, and carrier concentration [20, 21].

temperature [34, 35]. However, at temperatures below 300°C, we see the opposite effect [36]. While increasing the sintering temperature, we are removing more solvent, forming a ZnO structure, and increasing the grain size, which creates more pathways for conduction. As we further increase temperature, the grain size continues to increase—as seen in XRD—but more oxygen is being introduced to the ZnO. Increasing the grain size is expected to decrease the resistivity [37], while introducing more oxygen may increase the resistivity [38], resulting in a local minimum in the resistivity as a function of sintering temperature.

In general, doping, sintering, and post-processing all play a vital role in the conductivity of sol-gel ZnO films because of their effects on the ZnO lattice structure and defect formation. First, shallow donors can increase the free carriers in the conduction band. Second, the solvent must be completely evaporated, and the grain boundary concentration and adsorbed oxygen must be minimized to increase the mobility and carrier concentration, respectively. And lastly, post-processing techniques can be utilized to further improve the polycrystallinity and passivate defect charge states.

9. Photoconductivity

The MMR Hall effect system was equipped with a 365-nm light-emitting diode (LED), positioned 1.8 cm from the sample stage, to measure the resistivity as a function of light intensity. After dark measurements were taken, the LED light intensity was increased in steps up to 24 mW (~4.4×10^{16} photons·$cm^{-2} s^{-1}$), allowing the light and temperature to stabilize for at least 1 min prior to each measurement. Although light from the LED produces localized heating, the temperature was maintained at 300 K using a Joule-Thompson refrigerator located directly beneath the sample stage, operating in combination with a heating element. Photoconductivity was observed in IJP ZnO^5; IJP $IGZO^2$; and AJP ZnO^6, ZnO^7, and ZnO^8 (**Figure 11**).

Upon initial UV LED illumination at 0.98 mW (~4.4 × 10^{16} photons·cm^{-2} s^{-1}), there is a sharp decrease in resistivity. We credit this to ZnO absorbing light and promoting an electron from the valence band to the conduction band because the incident UV photons are of greater energy (~3.4 eV) than the ZnO bandgap (~3.2 eV). The photoresponse is due to oxygen chemisorption, where light illumination causes oxygen desorption and the release of trapped electrons to the conduction band [39]. Here, the greatest conductive response is seen at the greatest sintering temperature in AJP ZnO. This may be because larger grains desorb more oxygen when illuminated. In addition, the larger grain size would allow for better electron mobility.

With increased light intensity, IJP ZnO thin films quickly saturate, as there is no more oxygen to desorb. But, AJP ZnO—which has a greater photoconductive response—does not saturate at higher light intensity. As the intensity increases, photogenerated holes can be produced and then trapped at charged boundary states, while excess electrons can be promoted to the conduction band, increasing the free carrier concentration. In addition to the effect of the grain boundary, the charge state of defects may also undergo a change upon illumination and

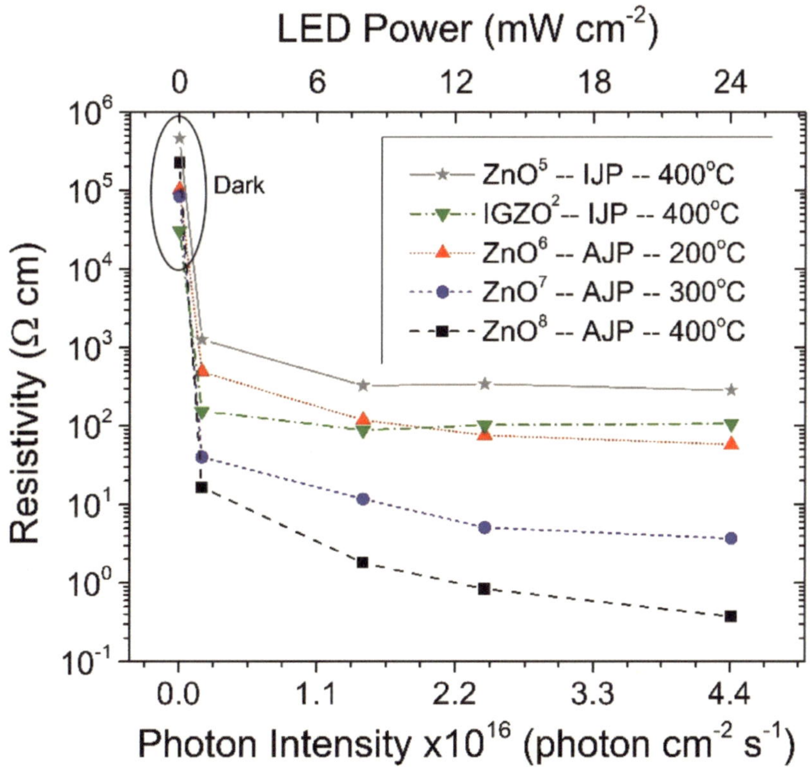

Figure 11. Resistivity measurements of printed ZnO films as a function of light intensity (365-nm LED) [20].

lead to an increase or a decrease in electron scattering affecting electron mobility. For instance, a change in the charge state of defects could increase electron scattering, decreasing the electron mobility and compromising the conductivity. Both the carrier concentration and electron mobility strongly affect the transport properties of ZnO films, and different photo-induced processes could lead to the observed non-linear behavior with increasing light intensity.

10. Positron annihilation spectroscopy

It is impossible to understand the effect of annealing on the transport properties without investigating the presence of point defects in the films. PAS is a well-established technique for measurements of cation vacancies, which strongly influence the transport properties [40–44]. In fact, many works have applied PAS and identified Zn vacancies in ZnO films and bulk single crystals [39–41]. The sensitivity of PAS to open volume defects such as vacancies can be understood as follows. The lack of positive ion cores at vacancies forms an attractive potential that traps positrons leading to characteristic changes in the measured positron annihilation parameters. Therefore, PAS is a very useful tool to further improve the development of sol-gel ZnO film. Here, depth-resolved Doppler broadening of PAS measurements was applied to elucidate the aforementioned effect of annealing on the electrical properties of ZnO films. The measurements were carried out on AZO films before and after annealing in forming gas (5% H_2, 95% N_2). **Figure 12a** and **b** shows S and W parameters, respectively, for the films as a function of incident positron energy and mean implantation depth. The S and W parameters represent the annihilation fraction of positrons with valence and core electrons, respectively, and they provide an indication about defect density [45–47]. The S parameter was obtained from the annihilation peak by dividing the counts in the central peak by the total counts in

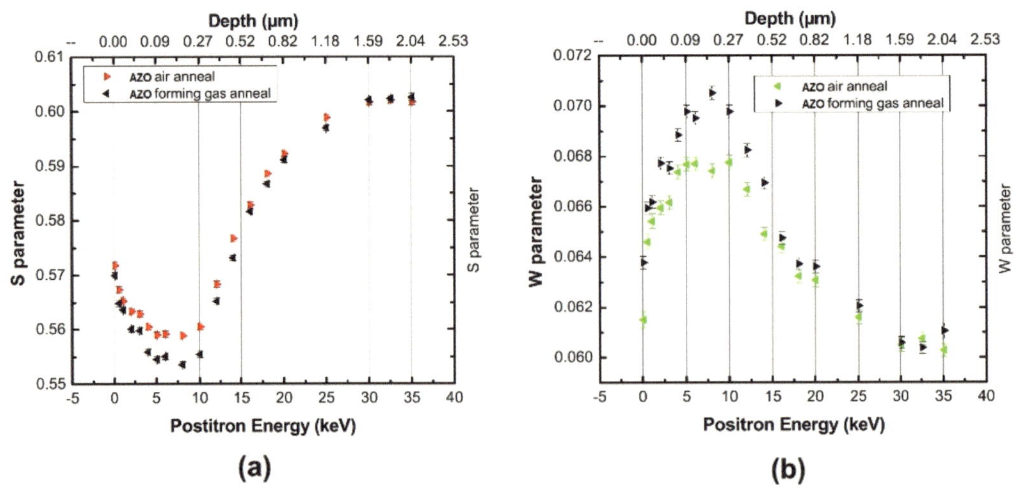

Figure 12. Depth-resolved PAS for AZO before and after forming gas post-processing: (a) S parameter as a function of positron beam energy and mean positron implantation depth and (b) W parameter as a function of positron beam energy and mean positron implantation depth [21].

the peak, while the W parameter was obtained by dividing the counts in the wings of the peak by the total counts in the peak. Trapped positrons at defects are more likely to annihilate with low-momentum valence electrons causing an increase in S parameter and a decrease in W parameter [45–47]. In **Figure 12a** and **b**, an increase in S parameter and a decrease in W parameter at low positron energy (0–5 keV) are due to positron annihilation at the surface of the films. The figures show a significant decrease in S parameter and an increase in W parameter after H_2 processing, which can be interpreted as follows. A Zn vacancy has a negative charge state and is therefore an effective trapping center for positrons, while an O vacancy or interstitial defects cannot trap positrons. Therefore, a decrease in S parameter and an increase in W parameter after H_2-annealing are a clear indication for the reduction or passivation of Zn vacancy-related defects. Annealing in forming gas cannot eliminate Zn vacancies; however, hydrogen can partially or completely fill Zn vacancies modifying their negative charge state, which prevents positron trapping. Similarly, Zn interstitials can fill Zn vacancies decreasing positron trapping. **Figure 13** shows the S-parameter versus W-parameter plot for ZnO films after air annealing, ZnO films after forming gas annealing, and ZnO bulk single crystals. The points on the graph represent the data points corresponding to energy values at which positrons annihilate only in the middle of the film without any influence from the surface or the substrate. The line in the S-parameter versus W-parameter plot runs through the bulk value, which is an indication that there is only one dominant defect type in the samples [40, 46]. This illustrates that H_2 annealing did not create new defects but only reduced Zn vacancies, the dominant trapping defect for positrons in ZnO. PAS studies here illustrated that Zn vacancy-related defects are dominant in sol-gel ZnO films and provided strong evidence that hydrogen passivates Zn vacancies, eliminating their deep acceptor state, which leads to a large increase in carrier concentration and high-induced conductivity in the films.

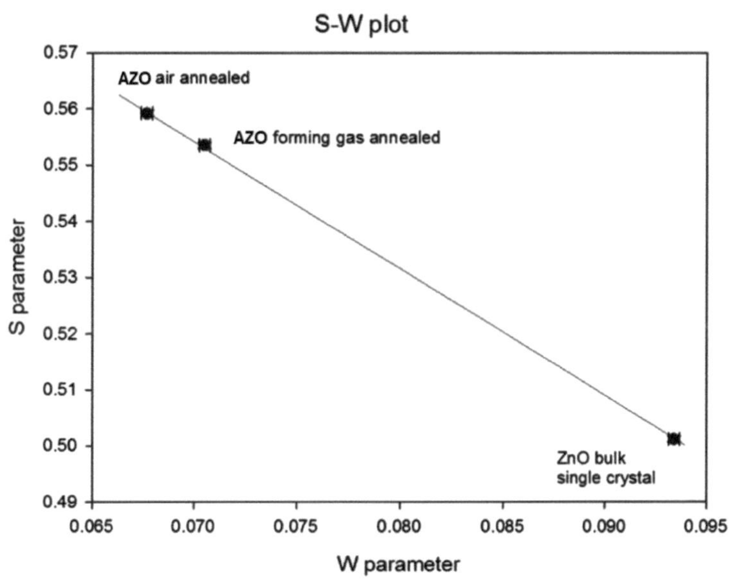

Figure 13. S-parameter versus W-parameter plot for bulk ZnO single crystal and AZO before and after processing in forming gas. The three points lie on a straight line, indicating one dominant defect type [21].

11. Conclusion

In conclusion, ZnO films were deposited on various substrates using a simple sol-gel precursor method. This precursor has proven compatible spin coating, IJP, and AJP techniques to fabricate TCOs and photodetectors. SEM measurements reveal surface roughness and non-uniformity that are inherent to the sol-gel process. However, these drawbacks can be overcome to optimize the UV-VIS and electrical properties. XRD analysis shows polycrystallinity that can be tuned by sintering temperature and processing atmosphere. The post-processing step and the addition of In^{3+} and Ga^{3+} have both shown to enhance the electrical conductivity of ZnO either through the suppression of acceptor vacancies or the addition of shallow donors. Resistive ZnO thin films also exhibited an overall photoconductive response of 10^6. PAS was executed to study the role of hydrogen passivation of cation vacancies in the electrical properties of sol-gel ZnO thin films and to illustrate its need for the development of conductive sol-gel ZnO films. Overall, this work demonstrates the compatibility of sol-gel ZnO with printed electronics and other devices and presents fundamental research to understand the structural, optical, and electrical properties of the material system.

Acknowledgements

The authors would like to thank the following collaborators for their contributive efforts: Wolfgang Anwand and Andreas Wagner at the Institute of Radiation Physics; Pooneh Saadatkia, Erik Flesburg, and Micah Haseman at Bowling Green State University; and Emily M. Heckman, Eric Kreit, Roberto S. Aga, Brett Wenner, Kevin Leedy, Steve Tetlak, David C. Look, Jeff Allen, and Monica Allen at the Air Force Research Laboratories at Wright-Patterson Air Force Base and Eglin Air Force Base.

Funding for this work was provided by multiple AFRL and DAGSI projects.

Conflict of interest

The authors declare that they have no conflicts of interest.

Author details

David Winarski[1,2] and Farida Selim[1,2*]

*Address all correspondence to: faselim@bgsu.edu

1 Department of Physics and Astronomy, Bowling Green State University, Bowling Green, Ohio, USA

2 Center for Photochemical Sciences, Bowling Green State University, Bowling Green, Ohio, USA

References

[1] Petti L, Münzenrieder N, Vogt C, Faber H, Büthe L, Cantarella G, et al. Metal oxide semiconductor thin-film transistors for flexible electronics. Applied Physics Reviews. 2016;**3**(2):021303

[2] Choi C-H, Lin L-Y, Cheng C-C, Chang C-H. Printed oxide thin film transistors: A mini review. ECS Journal of Solid State Science and Technology. 2015;**4**(4):3044-3051

[3] Smith M, Choi YS, Boughey C, Kar-Narayan S. Controlling and assessing the quality of aerosol jet printed features for large area and flexible electronics. Flexible and Printed Electronics. 2017;**2**(1):015004

[4] Teng L, Plötner M, Türke A, Adolphi B, Finn A, Kirchner R, et al. Nanoimprint assisted inkjet printing to fabricate sub-micron channel organic field effect transistors. Microelectronic Engineering. 2013;**110**:292-297

[5] Yin Z, Huang Y, Bu N, Wang X, Xiong Y. Inkjet printing for flexible electronics: Materials, processes and equipments. Chinese Science Bulletin. 2010;**55**(30):3383-3407

[6] Mohammed MG, Kramer R. All-printed flexible and stretchable electronics. Advanced Materials. 2017;**29**:1604965

[7] Liu Y-F, Tsai M-H, Pai Y-F, Hwang W-S. Control of droplet formation by operating waveform for inks with various viscosities in piezoelectric inkjet printing. Applied Physics A. 2013;**111**(2):509-516

[8] Eckstein R, Rödlmeier T, Glaser T, Valouch S, Mauer R, Lemmer U, et al. Aerosol-jet printed flexible organic photodiodes: Semi-transparent, color neutral, and highly efficient. Advanced Electronic Materials. 2015;**1**(8):1500101

[9] Aga RS, Lombardi JP, Bartsch CM, Heckman EM. Performance of a printed photodetector on a paper substrate. IEEE Photonics Technology Letters. 2014;**26**(3):305-308

[10] Lombardi JP, Bartsch CM, Aga RS, Heckman EM. Characterisation of DNA biopolymer-based UV photodetector fabricated by inkjet printing. Electronics Letters. 2015;**51**(10):778-780

[11] Girgis E, Fang M, Hassan E, Kathab N, Rao K. Ink-jet-printed $(ZnO)_{1-x}(TiO2)_x$ composite films for solar cell applications. Journal of Materials Research. 2012;**28**(03):502-506

[12] Liang YN, Lok BK, Wang L, Feng C, Lu ACW, Mei T, et al. Effects of the morphology of inkjet printed zinc oxide (ZnO) on thin film transistor performance and seeded ZnO nanorod growth. Thin Solid Films. 2013;**544**:509-514

[13] Oh S, Nagata T, Volk J, Wakayama Y. Nanoimprint for fabrication of highly ordered epitaxial ZnO nanorods on transparent conductive oxide films. Applied Physics Express. 2012;**5**(9):095003

[14] Meier M, Paetzold UW, Prömpers M, Merdzhanova T, Carius R, Gordijn A. UV nanoimprint for the replication of etched ZnO:Al textures applied in thin-film silicon solar cells. Progress in Photovoltaics: Research and Applications. 2013;**22**(12):1226-1236

[15] Hartner S, Khalil A, Ali M, Winterer M, Wiggers H. Ink-jet printed ZnO nanoparticles thin film for sensing applications. In: NSTI-Nanotech. Vol. 2. Boca Raton: CRC Press; 2010. pp. 535-537

[16] Brinker CJ, Scherer GW, editors. Sol–Gel Science: The Physics and Chemistry of Sol–Gel Processing. Saint Louis: Elsevier Science; 2014. p. 912

[17] Reisfeld R. Sol gel processed lasers. In: Sakka S, editor. Sol–Gel Technology (Handbook) 3. Dordrecht: Kluwer Anademic Publishers; 2004. pp. 239-261

[18] Reisfeld R. Doped polymeric system produced by sol–gel technology: Optical properties and potential industrial application. Polymer. 2005;**5**:95-103

[19] Reisfeld R, Saraidarov T. Innovative materials based on sol–gel technology. Optical Materials. 2006;**28**(1-2):64-70

[20] Winarski DJ, Kreit E, Heckman EM, Flesburg E, Haseman M, Aga RS, et al. Photoconductive ZnO Films Printed on Flexible Substrates by Inkjet and Aerosol Jet Techniques. Journal of Electronic Materials. 2017;**47**(2):949-954

[21] Winarski DJ, Anwand W, Wagner A, Saadatkia P, Selim FA, Allen M, et al. Induced conductivity in sol–gel ZnO films by passivation or elimination of Zn vacancies. AIP Advances. 2016;**6**(9):095004

[22] Mukhamedshina D, Mit' K, Chuchvaga N, Tokmoldin N. Fabrication and study of sol–gel ZnO films for use in Si-based heterojunction photovoltaic devices. Modern Electronic Materials. 2017;**3**(4):158-161

[23] Muslih EY, Munir B. Fabrication of ZnO thin film through chemical preparations. In: Ameen S, Akhtar MS, Shin H-S, editors. Emerging Solar Energy Materials. London: IntechOpen Limited; 2018. pp. 45-57

[24] Mahmood A, Naeem A. Sol-gel-derived doped ZnO thin films: Processing, properties, and applications. In: Chandra U, editor. Recent Applications in Sol–Gel Synthesis. London: IntechOpen Limited; 2017. pp. 169-193

[25] Znaidi L, Touam T, Vrel D, Souded N, Yahia S, Brinza O, et al. AZO thin films by sol–gel process for integrated optics. Coatings. 2013;**3**(3):126-139

[26] Liu Y, Li Y, Zeng H. ZnO-based transparent conductive thin films: Doping, performance, and processing. Journal of Nanomaterials. 2013;**2013**:196521

[27] Wang F-H, Chou C-T, Kang T-K, Huang C-C, Liu H-W, Kung C-Y. Structural, electrical, and optical properties of carbon nanotube-incorporated Al-doped zinc oxide thin films prepared by sol–gel method. Journal of Ceramic Processing Research. 2014;**14**:149-152

[28] Zhang X, Lee H, Kwon J-H, Kim E-J, Park J. Low-concentration indium doping in solution-processed zinc oxide films for thin-film transistors. Materials. 2017;**10**(8):880

[29] Khranovskyy V, Grossner U, Lazorenko V, Lashkarev G, Svensson BG, Yakimova R. PEMOCVD of ZnO thin films, doped by Ga and some of their properties. Superlattices and Microstructures. 2006;**39**(1-4):275-281

[30] Yadav AB, Periasamy C, Jit S. Study of post annealing effects on structural and optical properties of sol–gel derived ZnO thin films grown on n-Si substrate. In: Proceedings of IOP Conference Series: Materials Science and Engineering (ICMST '12); 10-14 June 2012; Kerala. New York: IOP Publishing; 2015. p. 012060

[31] López-Mena E, Jiménez-Sandoval S, Jiménez-Sandoval O. ZnO thin films prepared at low annealing temperatures, from a novel, simple sol–gel precursor solution. Journal of Sol-Gel Science and Technology. 2015;**74**(2):419-424

[32] Bel-Hadj-Tahar R, Mohamed AB. Sol–Gel processed indium-doped zinc oxide thin films and their electrical and optical properties. New Journal of Glass and Ceramics. 2014;**04**(04):55-65

[33] Nayak PK, Kim J, Chung S, Jeong J, Lee C, Hong Y. Spin-coated Ga-doped ZnO transparent conducting thin films for organic light-emitting diodes. Journal of Physics D: Applied Physics. 2009;**42**(13):139801

[34] Asghar M, Noor H, Awan M, Naseem S, Hasan M-A. Post-annealing modification in structural properties of ZnO thin films on p-type Si substrate deposited by evaporation. Materials Science in Semiconductor Processing. 2008;**11**(1):30-35

[35] Bouhssira N, Abed S, Tomasella E, Cellier J, Mosbah A, Aida M, et al. Influence of annealing temperature on the properties of ZnO thin films deposited by thermal evaporation. Applied Surface Science. 2006;**252**(15):5594-5597

[36] Shariffudin SS, Mamat MH, Herman SH, Rusop M. Influence of drying temperature on the structural, optical, and electrical properties of layer-by-layer ZnO nanoparticles seeded catalyst. Journal of Nanomaterials. 2012;**2012**:1-7

[37] Khan M, Bhatti K, Qindeel R, Alonizan N, Althobaiti HS. Characterizations of multilayer ZnO thin films deposited by sol–gel spin coating technique. Results in Physics. 2017;**7**:651-655

[38] Aghamalyan NR, Gambaryan IA, Goulanian EK, Hovsepyan RK, Kostanyan RB, Petrosyan SI, et al. Influence of thermal annealing on optical and electrical properties of ZnO films prepared by electron beam evaporation. Semiconductor Science and Technology. 2003;**18**(6):525-529

[39] Takahashi Y, Kanamori M, Kondoh A, Minoura H, Ohya Y. Photoconductivity of ultrathin zinc oxide films. Japanese Journal of Applied Physics. 1994;**33**(Part 1, No. 12A):6611-6615

[40] Krause-Rehberg R, Leipner HS. Positron Annihilation in Semiconductors: Defect Studies. Berlin: Springer; 1999. p. 383

[41] Selim FA, Winarski D, Varney CR, Tarun MC, Ji J, MD MC. Generation and characterization of point defects in $SrTiO_3$ and $Y_3Al_5O_{12}$. Results in Physics. 2015;**5**:28-31

[42] Čížek J, Valenta J, Hruška P, Melikhova O, Procházka I, Novotný M, et al. Origin of green luminescence in hydrothermally grown ZnO single crystals. Applied Physics Letters. 2015;**106**:251902

[43] Brillson LJ, Zhang Z, Doutt DR, Look DC, Svensson BG, Yu A. Interplay of dopants and native point defects in ZnO. Physica Status Solidi B. 2013;**250**:2110-2113

[44] Selim FA, Weber MH, Solodovnikov D, Lynn KG. Nature of native defects in ZnO. Physical Review Letters. 2007;**99**(8):085502

[45] Hautojäervi P. Positrons in Solids. Heidelberg: Springer; 1979. p. 258

[46] Schultz P, Lynn KG. Interaction of positron beams with surfaces, thin films, and interfaces. Reviews of Modern Physics. 1988;**60**(3):701-779

[47] Selim FA, Wells DP, Harmon JF, Williams J. Development of accelerator-based γ-ray-induced positron annihilation spectroscopy technique. Journal of Applied Physics. 2005;**97**:113539

Chapter 6

Sol-Gel Films: Corrosion Protection Coating for Aluminium Alloy

Evelyn Gonzalez, Nelson Vejar, Roberto Solis,
Lisa Muñoz, Maria Victoria Encinas and
Maritza Paez

Additional information is available at the end of the chapter

http://dx.doi.org/10.5772/intechopen.79712

Abstract

Aluminum alloys used in aeronautical industry are susceptible to corrosion. The solution to this problem is base chromate materials, which have been heavily regulated and restricted. The development of alternatives begins in the 1970s and the 2000s, where some potential methodologies were established. The sol-gel process is one of these methods, in which thin oxide layers are deposited on the metal substrate. An important aspect is the fact of possible combinations among types of oxides and the incorporation of an organic compound to improve the performance of the films; moreover, this allows the addition of inhibitors and nanomaterials, making this method an interesting and versatile way to obtain a coating. In this chapter, we will describe the importance of the use of coating synthesized via sol-gel in the corrosion protection of metal surfaces. The advantages and disadvantages of using modified sol-gel polymer films and hybrid system coatings will also be discussed, as well as the methodologies for the chemical characterization and the feasibility of evaluating the mechanical properties of the coatings.

Keywords: sol-gel coating, aluminum alloy, hybrid, corrosion, biocorrosion

1. Introduction

The 2xxx aluminum alloys are widely used in the aircraft industry due to their high specific strength and lightweight [1]. These alloys contain elements, as copper, used to improve their mechanical properties. The presence of this element, together with others of lower content, and the history of associated thermal treatments, promotes the formation of some copper-rich sites, known as intermetallics. Unfortunately, the heterogeneous microstructures of intermetallic make 2024 alloy become more susceptible to pitting corrosion in the media containing chloride ions, due to the formation of microscopic galvanic couples [2].

Metallic corrosion occurs because of chemical reactions between the metal surface and the environment, changing the metal over its original ore. To prevent the beginning of localized corrosion processes and to extend the service life, in the aircraft industry, the most common practice is to avoid the direct contact of the electrochemically active matrix with the surrounding environment by applying a protective coating system [3].

The traditional surface passivation treatment for aluminum alloy is conversion coating, which is produced in two steps: (i) dissolution of the base metal through reaction with the passivating solution and (ii) precipitation of insoluble compounds, a layer of corrosion product capable of resisting further chemical attack [4]. Chromate conversion coatings, typically generated from mixtures of soluble hexavalent chromium salts and chromic acid, participate in oxidation-reduction reactions with aluminum surfaces, precipitating a continuous layer of insoluble trivalent chromium and soluble hexavalent chromium compounds [5].

Corrosion protection occurs as hexavalent chromium leaches into defect sites, forming dense, insoluble trivalent chromium products. Chromate conversion coatings comparatively promote very good adhesion of organic coatings and offer as a whole system excellent corrosion protection [6]. The hexavalent chromium-containing compounds used in chromate conversion coatings are known to be carcinogenic and generally regarded as very hazardous soil and groundwater pollutants. Stricter environmental regulations have mandated the near-term removal of Cr(III)-containing compounds from corrosion inhibiting packages used for the protection of aluminum-skinned aircraft. Therefore, the need for the development of protection process exists, following nontoxic, chromium-free and environmentally friendly materials and protocols.

Several techniques are used for the deposition of coatings on metals; these methods include physical vapor deposition (PVD), chemical vapor deposition (CVD), electrochemical deposition, plasma spraying, and sol-gel process. The sol-gel process has emerged as a versatile method for preparing a host of oxide materials to protect the metal surface [7]; moreover, sol-gel materials are candidates as it is possible to form highly adherent, chemically inert films. In comparison with other deposition technologies, sol-gel technics offer several potential advantages, such as (i) preparation in room temperature, (ii) diverse and complex system, (iii) cured treatment at relative low temperature, and (iv) considered as a "green method" [8]. Thin films may be readily prepared from water-based systems, resulting in low volatile organic compound (VOC) content materials and processes. Instead, the primers and topcoats have VOC contents of 340 and 420 g/l, respectively, in comparison with the aqueous sol-gel solutions suitable for spray coating on aluminum substrates, which have a VOC content of 100–200 g/l [9]. On the other hand, the method allows to obtain thin films of sub-micrometer thickness with high purity in multiple combinations. By forming dense coatings, sol-gel films act as barriers for diffusion of aggressive species, such as chlorine and oxygen, blocking the electron transfer of metal surface to and from the environment. Moreover, the flexibility of the sol-gel process also permits the incorporation of corrosion inhibiting compounds, thereby providing another mechanism for corrosion protection. These characteristics lead to the possibility of forming environmentally compliant coatings capable of improving corrosion resistance without the use of metal chromates or the generation of liquid hazardous waste products [6].

The anticorrosion behavior of coating is studied using electrochemical methods, which allow to obtain the susceptibility of metal to be corroded. The most important technics used are polarization curves and electrochemical impedance spectroscopy.

In the following topics, we will describe the methodology to obtain coating from oxide species using sol-gel technics. Moreover, the diversity and complex system of hybrid coating will be reviewed. In this way, the advantages and disadvantages of using modified sol-gel polymer films for the generation of smart coatings will be discussed also. Finally, the chemical characterization and the feasibility of evaluating the mechanical properties of the coatings will be analyzed as well.

2. Synthesis and deposition of sol-gel coatings

The sol-gel process can be described as the evolution of an oxide network by continuous condensation reactions of molecular precursors in a liquid medium [10]. Two ways to prepare sol-gel coating have been proposed: the inorganic method and the organic method. The inorganic method involves the evolution of networks through the formation of a colloidal suspension (usually oxides) and gelation of the sol (colloidal suspension of very small particles (1–100 nm)) to form a network in continuous liquid phase. But the most widely used method is the organic approach, which generally starts with a solution of metal/metalloid alkoxide precursors, M(OR)n, in an alcohol or other low molecular weight organic solvent, where M can represent different elements such as Si, Ti, Zr, Al, Fe, B, etc. and R is typically an alkyl/allyl group. Sol-gel processing proceeds in several steps which will be discussed later: (i) hydrolysis and condensation of the molecular precursors and formation of sols, (ii) gelation (sol-gel transition), (iii) aging, and (iv) drying [11].

2.1. Hydrolysis and condensation

In the sol-gel process, hydrolysis and condensation are equilibrium reactions and can proceed simultaneously once the hydrolysis reaction has initiated. The reaction mechanisms for acid or base catalysis are very different and have to be considered separately [12]. The pH is an especially important parameter to control the morphology of coatings. At intermediate pH, the reaction rate of condensation is proportional to the concentration of the OH^- ions. At pH lower than about 2, the silicic acid species are positively charged, and the reaction rate of the condensation is proportional to the concentration of H^+. While under strong alkaline conditions, the solutions contain mainly anionic species. For this reason, the rate of Si–O–Si cleavage or redissolution of particles is high at alkaline pH.

Under acidic conditions, the oxygen atom of a ≡Si–O^-, ≡Si–OH, or ≡Si–OR group is protonated in a rapid first step. A good leaving group (water or alcohol) is thus created. In addition, electron density is withdrawn from the central silicon atom, rendering it more electrophilic and thus more susceptible to attack by water (in hydrolysis reactions) or silanol groups (in condensation reactions).

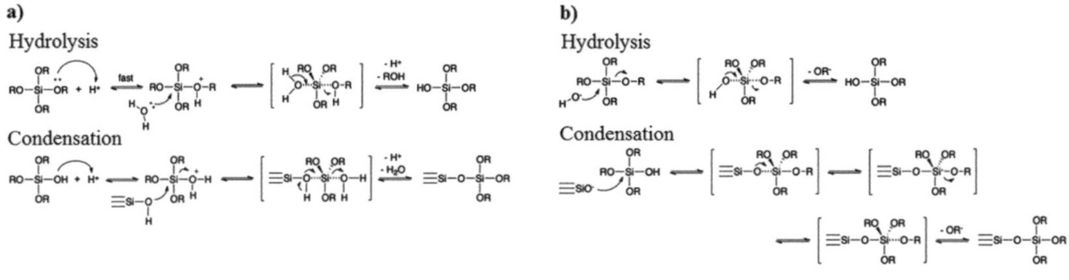

Figure 1. General mechanisms of synthesis sol-gel catalyzed (a) acid, and (b) base conditions.

Under basic conditions (**Figure 1**), the reaction proceeds by nucleophilic attack of either an OH^- (in hydrolysis reactions) or a ≡Si—O^- ion (in condensation reactions) to the silicon atom with an SN_2-type mechanism. The entering OH^- or ≡Si—O^- group is formed by deprotonation of water or a ≡Si—OH group. Under strong alkaline conditions, the Si—O—Si bonds can be cleaved again by OH^-. Inductive effects of the substituents attached to a silicon atom are very important, because they stabilize or destabilize the transition states or intermediates during hydrolysis and condensation. The electron density at the silicon atom decreases in the following order: ≡Si—R\` > ≡Si—OR > ≡Si—OH > ≡Si—O—Si.

For acid catalysis, the electron density at the silicon atom should be high since the positive charge of the transition state is then stabilized better. Therefore, the reaction rates for hydrolysis and condensation under acidic conditions increase in the same order as the electron density. For base catalysis, a negatively charged intermediate must be stabilized.

Therefore, the reaction rates for hydrolysis and condensation increase in the reverse order of the electron density.

2.1.1. Gelation, aging, and drying

During the gelation, the colloidal particles and condensed species link together to become a three-dimensional network and the viscosity increases sharply. Physical characteristics of the gel network will depend greatly upon the size of particles and extent of cross-linking prior to gelation [13]. Aging of the prepared sol-gel prior to application on the metallic substrate also affects strongly the corrosion protection properties of the resulting coatings. Aging of the sol can promote the condensation reactions of the precursors, including formation of further crosslinks and increasing the viscosity of the sol-gel, which can eventually lead to the formation of thick coating with a high defect density [14]. During drying, loss of water, alcohol, and other volatile components takes place. The evaporation of the liquid from a wet gel generally proceeds in more than one stage, where the liquids flow through the polymer evolving to a stable rigid condition, and where the effect of the surface tension on the mechanical properties of the final coating, is also considered [15].

Two processes are important for the collapse of the network. First, the slower shrinkage of the network in the interior of the gel body results in a pressure gradient that causes crack. Second, larger pores will empty faster than smaller pores during drying; that is, if pores with different radii are present, the meniscus of the liquid drops faster in larger pores. The wall between pores of different sizes is therefore subjected to uneven stress and crack. Low-temperature drying is normally employed for drying of hybrid sol-gel coatings entrapping organic compounds. Although compact crack-free films can be obtained, room temperature cured sol-gel coatings exhibit higher water sensitivity compared to those cured at higher temperatures. Higher cure temperatures (up to 200°C) promoting condensation reactions and formation of dense hybrid coating improve the barrier properties. By controlling the aging and drying conditions, further pore size and mechanical strength control may be achieved.

Despite the fact that the most used alkoxides are the silicon-type in sol-gel coating synthesis, it is pertinent to make a brief mention about the transition metal alkoxides.

There are two important differences between silicon and transition metal alkoxides that have to be considered when we want to synthesize a sol-gel coating [16]: (i) metals are more electropositive (Lewis acidic) than silicon and therefore more susceptible to a nucleophilic attack and (ii) the preferred coordination number is higher than their valence. The increase of the coordination number beyond the valence is reached by interaction with any nucleophilic (Lewis basic) entity in the system. When we compare SiO_2 and TiO_2, both central atoms are in the IV oxidation state. However, silicon is always four coordinated (tetrahedral SiO_4 building blocks) while titanium in rutile is six coordinated (octahedral TiO_6 building blocks). The mechanisms of condensation reactions of metal alkoxides are similar to those of silicon alkoxides in a sense that an M—OH group undergoes nucleophilic attack by another metal atom. Due to the higher propensity of metal atoms to interact with nucleophilic agents, base or acid catalysts are not needed in most cases.

When a silica network grows, the question that decides the morphology of the obtained coating is whether condensation occurs preferentially at the end of chain of corner sharing SiO_4 tetrahedra or at a central atom. For transition metals, this issue is more complicated and hardly understood in detail in most cases. An additional difference between metal alkoxide and silicon alkoxide-driven sol-gel process is the morphology of the final material. While in the silicon-based sol-gel process only amorphous materials are produced, the metal alkoxides can form crystalline compounds.

2.1.2. Considerations

Two points are considered by the time of synthesis of sol-gel coating, alkoxy group/H_2O ratio (Rw) and solvent. Alkoxides are employed as precursors for the sol-gel process, as mentioned above. In the case of silicon, the most prominent alkoxides are tetramethoxysilane (TMOS) and tetraethoxysilane (TEOS) (**Figure 2**) [10]. Both precursors are liquid under standard conditions, and TMOS shows a faster hydrolysis reaction compared to TEOS but, at the same time, generates methanol, which is avoided for its toxicity. The application of these precursors in the sol-gel process would lead to a three-dimensional network and finally, after heating, to a coating. Considering that alkoxides must first be hydrolyzed before condensation reactions can take place, the hydrolysis rates of alkoxysilanes are influenced by both the inductive effects and steric factors. Any branching of the alkoxo group or increasing of the chain length lowers the hydrolysis rate of the alkoxysilanes. It means that the reaction rate decreases in the order $Si(OMe)_4$ > $Si(OEt)_4$ > Si(O``Pr)$_4$ > Si(O`Pr)$_4$.

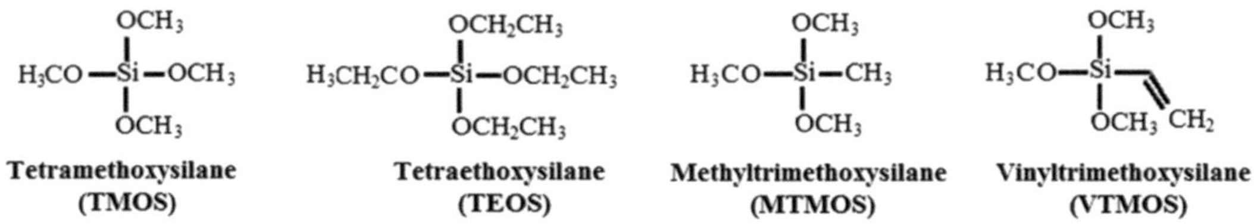

Figure 2. Examples of some precursors commonly used in sol-gel coatings.

The overall reaction for sol-gel processing of tetraalkoxysilanes implies that two equivalents of water (Rw = 2) are needed to covert $Si(OR)_4$ if no condensation takes place. Increasing the water proportion generally favors the formation of silanol groups over Si—O—Si groups. The Rw, together with the kind of catalyst, strongly influences the properties of the silica gels [17].

A solvent may be necessary to homogenize the reaction mixture of alkoxide-based systems, especially at the beginning of the reaction. Polarity, dipole moment, viscosity, and protic or non-protic behavior of the solvent influence the reaction rates and thus the structure of the sol-gel coating. Polar and particularly protic solvents (H_2O, alcohols, etc.) stabilize polar species such as ($Si(OR)_x(OH)_y$)n by hydrogen bridges. The latter generally play a very important role in sol-gel systems. Nonpolar solvents (dioxane and tetrahydrofuran) are sometimes used for organotrialkoxysilanes (R`$Si(OR)_3$) or incompletely hydrolyzed alkoxide systems [17].

2.2. Application techniques of the sol-gel coatings

A sol-gel coating can be applied to a metal substrate through various techniques, such as dip coating and spin coating, which are the two most commonly used coating methods. Spraying [18] and electrodeposition [19] also emerged recently and could be the major sol-gel coating application methods in the future. In both methods, spin coating and dip coating, the sol-gel is directly deposited onto the support. The condensation reaction can also occur between silanol and hydroxyl groups of the metal (obtained by the activation of the surface with bases), leading to the covalent bonding of silane to the surface: —SiOH + HO-surface → —Si—O-surface + H_2O (1).

It is generally accepted that during the sol-gel process, the sol precursor first hydrolyzes, and then, the hydrolyzed species are adsorbed onto the surface undergoing cross-linking to form a continuous film.

2.2.1. Spin and dip coating

The production of thin films by spin coating was initially reported by Ogawa in the 1996 [20]. Among other techniques, spin coating is the most easily applicable one for obtaining uniform thin layers on flat surfaces [21]. Experimentally, a small amount of the coating material is deposited onto the center of the support. Subsequently, the support is rotated at high speed in order to spread the coating material by the centrifugal force. In general, the higher the rotation speed, the thinner the film. Therefore, by selecting the appropriate spin rate, it is possible to modulate the film thickness [21].

Besides the very interesting results obtained and the very good performances of this simple deposition technique, spin coating possesses some drawbacks concerning the size and shape of the substrates. In fact, as reported by Tyona [22], large supports are difficult to be homogeneously deposited by this method. Additionally, in a typical spin-coating deposition, minority of the 5% of the starting solution is deposited successfully onto the substrate forming the thin film, whereas the complement percentage is lost due to the rotation of the spinner. Further, the final morphology of the coated substrate can be influenced by several parameters such as spin speed, time of spin, acceleration, fume exhaust, etc.

Dip coating is one of the most convenient methods used in the laboratory and industry to deposit films onto a metallic surface with a controlled thickness from a sol-gel solution. This method is simple and provides excellent reproducibility [23]. Basically, the method may be separated into three important stages: (1) *Immersion and dwell time*: The substrate is immersed into the precursor solution at a constant speed followed by a certain dwell time in order to leave sufficient interaction time of the substrate with the coating solution for complete wetting. (2) *Deposition and drainage*: By pulling the substrate upward at a constant speed, a thin layer of precursor solution is entrained. Excess liquid will drain from the surface. (3) *Evaporation*: The solvent evaporates from the fluid, forming the as-deposited thin film, which can be promoted by heated drying. Subsequently the coating may be subjected to further heat treatment to obtain a more dense film [10].

2.2.2. Electrochemical deposition

The electrochemical deposition of sol-gel films provides an alternative for shifting the pH on the substrate. In aerated aqueous media, it is well known that by applying cathodic potential, the following reactions occur at the electrode surface:

$$O_2 + 2\,H_2O + 4e \rightarrow 4\,OH^- \tag{1}$$

$$2\,H_2O + 2e \rightarrow 2\,OH^- + H_2 \tag{2}$$

Both reactions generate OH^- ions that increase the interfacial pH near the cathode, which catalyzes the sol-gel process facilitating the film formation. There are three advantages of electrodeposition technic: (1) pH varies only close to the cathode, so the stability of the bulk solution is not affected, (2) the deposition process is controllable by electrochemical parameters, and (3) the film deposition is restricted to the conducting part of the surface and controlled by the kinetics of the electrochemical process [24].

2.3. Hybrid organic-inorganic sol-gel coatings

In order to overcome the limitations associated with conventional inorganic sol-gel coatings, such as brittle oxide films, thicker coatings (>1 μm), crack-free, and relatively high temperatures (400–800°C), the hybrid coatings by the incorporation of organic groups in the inorganic sol-gel network have shown good results [24].

Two different approaches can be used for the incorporation of organic groups into an inorganic network by sol-gel processing, namely, embedding of organic molecules into gels without chemical bonding (class I hybrid materials) and incorporation of organic groups through covalent bonding to the gel network (class II hybrid materials). Embedding of organic molecules is achieved by dissolving them in the precursor solution. The gel matrix is formed around them and traps them, and the organic and inorganic entities interact only weakly with each other. The inorganic network and the organic network interpenetrate but are not bonded to each other [14]. Despite the presence of weak dispersion forces and Van der Waals interactions between organic and inorganic components of such hybrids, the physical bonds are not stable enough for long-term applications involving weathering. Formation of strong covalent bonds between organic and inorganic components can significantly improve corrosion protective properties of the hybrid coatings. Very important sol-gel materials are obtained when functional or nonfunctional organic groups are covalently linked to oxide networks (class II hybrid materials). Silicate hybrids are mostly done by using organotrialkoxysilanes, $R`Si(OR)_3$, as precursors to sol-gel processing. Nearly any organic group R` can be employed; the only requirement is that the group R` must be hydrolytically stable. Since Si—C bonds are hydrolytically stable, the organic groups are retained in the final material after sol-gel processing.

Different functional groups impart different corrosion protective properties to hybrid coatings. Moreover, the corrosion protective properties of the hybrid coatings dramatically depend on the presence, the type, and the number of the reactive groups of the used agent. For this reason, organotrialkoxysilanes are typically copolymerized with tetraalkoxysilanes or metal alkoxides to obtain the properties characteristic of highly cross-linked networks. This allows incorporation of organic groups without lowering the network connectivity because one Si—O—Si entity is replaced by Si—R``—Si. The groups R`` can range from simple alkylene or arylene groups to more complex entities. The hybrid sol-gel coatings containing functional groups show a higher cross-link density and better mechanical properties [25]. Not only the nature of organic components but also their content in the hybrid sol-gels plays a very important role in the final properties of the hybrid coatings. An increase in the organic content of the hybrid coatings leads to the formation of less porous and thicker films appropriate for barrier protection of metals. However, a high concentration of organic component can lower the adhesion and the mechanical properties of the final coating. So it is important to point out that there is an optimum ratio for inorganic-organic components to deliver maximum corrosion resistance. The optimum organic/inorganic ratio varies depending on the precursors employed and on the coating application technique [26]. Hydrophobic hybrid coatings can reduce the kinetics of the corrosion processes by delaying penetration of water and other electrolytes toward the metal/coating interface. However, a prolonged exposure of the hybrid coatings to water/electrolyte will eventually result in moisture penetration of the metal/coating interface. Considering the reversible nature of hydrolysis and condensation reactions involved in the creation of the coating, water penetration can promote hydrolysis of the bonds formed during condensation reaction resulting in delamination [27]. The final film can carry specific organic functions, which can present certain properties, such as good adhesion, self-healing, abrasion resistance, scratch resistance, hydrophobicity, etc. Network formation is only possible if the precursor used has at least three possible cross-linking sites. Both, tetraalkoxysilanes $Si(OR)_4$ and trialkoxysilanes $(RO)_3SiR'$, possess this ability.

2.3.1. *Sol-gel coatings synthesized in the presence of a performed polymer*

This route of synthesis means that the polymer is mixed with the precursors and hydrolysis and condensation are started [28]. It is important to determine the best reaction conditions to avoid phase's separation. Therefore, the right choice of solvent is of major significance. Typical polymer solvents depending on the functional groups and polarity of the polymers are tetrahydrofuran (THF), dimethoxyethane (DME), alcohols (methanol, ethanol, isopropanol, etc.), acetic acid, etc. During sol-gel reaction, alcohols are liberated which can change the solvent properties resulting to precipitation of the initially soluble polymers, leading to heterogeneous films. Therefore, the choice of the polymer and solvent is important in this synthetic route. Polymers with functional groups can interact with the sol-gel structures, for example, by hydrogen bonding, such as alcohols or amines. In many cases, an effective interaction between the polymer and the inorganic structure results in a homogeneous distribution of small inorganic structures in the polymer matrix.

2.4. Doping of the sol-gel coatings

Despite the effective barrier protection of metallic substrates by hybrid sol-gel coatings, these systems are prone to fail because of water ingress into the films. For this reason, incorporation of active species such as binding agents and corrosion inhibitors, which add active protection mechanisms to the system, can improve the protective properties of the hybrid sol-gel coatings. Thus, incorporation of nanoparticles such as silica, ceria, zirconia, alumina, titania, and zeolite, as mechanical reinforcement, were the first proposed approaches for modification of hybrid sol-gel coatings [29]. The improved mechanical properties, increased thickness, and lower crack sensitivity achieved by addition of a controlled amount of the particles resulted in enhanced corrosion protection of the underlying substrate. However, the particle size and surface modification have shown to be critical, as agglomeration of the embedded particles promoted by gelation process could lead to coating rupture and deterioration of the coating barrier properties [14]. It is important to point out that the critical dopant concentration, which physical/mechanical properties in the coating starts to degrade, must always be considered. Moreover, a strong interaction between particle and matrix interfaces is required. Corrosion inhibitors can either be added (i) directly to the coating formulation or (ii) immobilized in carriers to reduce the possible interactions with the matrix and control release of the inhibitor [30]. In addition, nanoparticles not only can be added but also can be formed in situ in the coatings, eliminating some of the challenges associated with the strong interfacial forces between matrix and particles [31].

2.4.1. *Direct and indirect addition of inhibitors*

The most common way of inclusion of corrosion inhibitors into sol-gel systems is mixing them with the coating formulation [32]. The most important factor to be considered in such systems is the solubility of inhibitor in the corrosive media. While a low solubility of inhibitor can lead to a weak self-healing effect due to the low concentration of active agents at damaged site, a high solubility will limit prolonged healing effect because of rapid leach out of the active agents from coating, producing the coating degradation by blistering and delamination. Despite the potential drawback of this class of extrinsic self-healing sol-gel coatings, they have been extensively studied for protection of different metallic substrates due to ease of preparation. The corrosion inhibitors used can be divided according to their nature into (i) inorganic and (ii) organic inhibitors [33]. Some of the most used inorganic inhibitors are the rare earth metals and some transition metals such as Ce, La, and Zr which have showed an improved anticorrosive performance in the doped hybrid coatings compared to the undoped ones [34]. Incorporation of the active Ce ions not only facilitates preparation of dense and defect-free hybrid coatings but also increases the protection mechanism via selective leaching of Ce ions to the damage site restoring the coating's protective properties [35]. Organic inhibitors prevent corrosion by either increasing the anodic or cathodic polarization resistance of the corrosion cell or retarding diffusion of corrosive agents to the metallic surface [36]. However, their inhibition efficiency depends on the chemical composition, molecular structure, and affinity of the metal surface. Organic inhibitors such as phosphonic acid, 2-mercaptobenzothiazole (MBT), 2-mercaptobenzimidazole (MBI), benzotriazole (BTA), etc. have been successfully incorporated into sol-gel systems to improve their corrosion protection properties by inducing active protection [37]. In several cases, release of organic molecular species from the hybrid sol-gel matrix is based on a pH-triggered release mechanism. With this method, it is possible to release inhibitors only at damaged areas due to local pH changes.

Although incorporation of corrosion inhibitors into sol-gel coatings is a promising route in the development of active corrosion protective hybrid coatings, there are inevitable drawbacks associated with direct mixing of active agents into coating formulation. Firstly, it is quite difficult to control leach out of entrapped inhibitors especially when they are poorly soluble within the coating matrix. Secondly, inhibitors can chemically interact with the coating matrix losing their own activity and lowering the barrier properties of the matrix. A probable solution to this problem is the encapsulation of active species or complexing them with other chemicals [38]. A quite simple approach for inhibitor entrapment/immobilization is based on the complexation of organic molecules with β-cyclodextrin. Cyclodextrins are cyclic oligosaccharides that possess a unique molecular cup-shaped structure with a hydrophilic exterior and a hydrophobic interior cavity. They are able to form complexes with various organic guest molecules which fit within their cavities. Organic aromatic and heterocyclic compounds are normally the main candidates for the inclusion complexation reaction. 2-Mercaptobenzothiazole (MBT) and 2-mercaptobenzimidazole (MBI) were successfully loaded in β-cyclodextrin [33]. In the case of cyclodextrin complexes, incorporation of the inhibitor-loaded particles in sol-gel coatings has been more efficient than direct inhibitor loading in imparting long-term self-healing function. On the other hand, ceramic particles such as silica and alumina can be employed as micro-/nano-containers to immobilize corrosion inhibitors. The selected inhibitors can be entrapped on the carriers through controlled hydrolysis of the relevant precursors in the inhibitor-containing aqueous solutions [39].

3. Physical-chemical characterization

The proposal of this topic is to show an overview of some methodologies of characterization in order to understand the information related to properties of film coating. The most useful and used technics to characterize the sol-gel coating are infrared spectroscopy (IR), X-ray photoelectron spectroscopy (XPS), and scanning electron microscopy (SEM).

3.1. Infrared spectroscopy (IR)

Infrared spectroscopy is based on the vibrations of atoms of a molecule. An IR spectrum is obtained by passing an IR radiation through a sample and determining what fraction of the incident radiation is absorbed at determinate energy. The energy at which any peak in an absorption spectrum appears corresponds to the frequency of a vibration of a part of a sample molecule. The interactions of IR radiation with matter may be understood in terms of changes in molecular dipoles associated with vibrations and rotations [40].

IR technic allows characterize bonds Si—O, Si—Si, and Si—C. Furthermore, this analysis is used to determine the presence of active molecules in hybrid sol-gel film which has been modified using, among others, organic substituent such as hydrocarbon chain (C—H, C—C) [41], organic compound [42], and inhibitor [43].

3.2. X-ray photoelectron spectroscopy (XPS)

X-ray photoelectron spectroscopy is an established quantitative method for the determination of elemental abundance and the assessment of chemical binding [44]. Photoelectron production in its simplest form describes a single-step process in which an electron initially bound to an atom/ion is ejected by a photon. Since photons are a massless (zero rest mass), charge less package of energy, these are annihilated during photon-electron interaction with complete energy transfer occurring. The general equation for this process is $h\upsilon = BE + KE + \varphi_{spec}$. If this energy is sufficient, it will result in the emission of the electron from the atom/ion as well as the solid. The kinetic energy (KE) that remains on the emitted electron is the quantity measured. This is useful since this is of a discrete nature and is a function of the electron binding energy (BE), which, in turn, is element and environment specific, and φspec is the work function of the electron spectrometer, which is usually quite small (< 5 eV) compared to BE and KE [45]. It is convenient in surface analysis to measure BE and KE with respect to the Fermi level. Since the binding energies of core electrons are different in different atoms, XPS is capable to identify the elemental compositions of materials by measuring the KEs of their ejected electrons. XPS can detect all the elements except for H and He. In addition, XPS is sensitive to the chemical environments of the atoms it detects.

The popularity of XPS stems from its ability to: (a) Identify and quantify the elemental composition of the outer 10 nm or less of any solid surface with all elements from Li-U detectable. Note: This is on the assumption that the element of interest exists at >0.05 atomic %. (b) Reveal the chemical environment where the respective element exists in, that is, the speciation of the respective elements observed. (c) Obtain the information above with relative ease and minimal sample preparation [45].

In this way, the XPS help in the analysis of sol-gel coating in order to determine the oxide state of doped polymer [46], the presence of metal [47], and the bond between metal and polymer [6].

3.3. Scanning electron microscopy (SEM)

A basic SEM consists of an electron gun (field emission type or others) that produces the electron beams; electromagnetic optics guide the beam and focus it. The detectors collect the electrons that come from the sample (either direct scattering or emitted from the sample), and the energy of the detected electron together with their intensity (number density) and location of emission is used to put the image together. SEM also offer energy dispersive photon detectors that provide analysis of X-rays that are emitted from the specimen due to the interactions of incident electrons with the atoms of the sample [48].

SEM technic allows to characterize the coated metallic surface [49] and determine the thickness of deposited polymer [41].

4. Mechanical characterization

The properties of sol-gel coatings have a strong dependence on the substrate on which they have been generated. The requirements for the coating vary depending on the type of substrate, ranging from purely physical (e.g., optical properties), through chemical (e.g., anticorrosion properties), to purely mechanical (e.g., resistance to wear). The type of coating generated is a direct function of the desired final properties, being able to choose between inorganic coatings or hybrid coatings (organic-inorganic).

The main qualities required of any coating generated on a metallic substrate, regardless of its application in service, are:

- Homogeneity of the thickness of the obtained coating
- Homogeneity in the chemical composition of the coating, to present the same mechanical behavior throughout the sample
- High adhesion to the substrate, guaranteeing structural and mechanical stability over time

The first two requirements are easily achievable with sol-gel coatings. Regarding the adhesion, ceramic coatings obtained following the sol-gel route present a high adhesion to the metallic substrates due to the presence of hydroxyl radicals (—OH) on the surface of the latter which manage to form a chemical bond between atoms of the deposited *gel* and atoms of the outer surface of the substrate.

Aluminum alloy substrates have been coated with sol-gel to improve their corrosion behavior, using mainly alloys with aeronautical or automotive applications, as well as structural interest in the civil field. The surface preparation of the substrates to be coated is usually initiated with chemical degreasing. Subsequently, the substrate can be simply coated, or it can be subjected to the generation of a certain roughness by roughing or polishing. The coating generated by the sol-gel route can be the only protection system, or it can be used in combination with other systems, such as special paint for aeronautical applications.

The main characteristics of the sol-gel coatings on aeronautical aluminum alloys are anticorrosive and mechanical together with wear behavior.

García-Heras et al. [50] demonstrate the importance of the surface preparation of the substrate and the concentration of alkoxide precursor in the anticorrosive efficiency of silica coatings manufactured on the 2024 T6 aluminum alloy. Hamdy and Butt [51] demonstrate the effectiveness against corrosion of inorganic silica coatings, starting from TEOS as a precursor, on the 6063 aluminum alloy without anodizing and anodizing prior to deposition, as well as the influence of the treatment thermal densification. It has been reported that the use of hybrid coatings generates, on the one hand, greater coating thicknesses, and in addition, a very adherent surface is formed for the subsequent painting system on the 2024 alloy [52]. The same degree of protection as by coating and painting systems has been achieved by Liu et al. [53]. An alternative way of generating hybrid coatings is by adding inorganic particles to sol solutions of alkoxides with non-hydrolyzable groups [54], although the percentage of added particles must be optimized, since an excess means the formation of thicker coatings but with pores, favoring the formation of pitting corrosion. The amount of inorganic particles added to the sol-gel is not the only determining factor to obtain a good behavior against corrosion, since hydrophobic particles generate greater resistance to corrosion than hydrophilic particles [55].

It should be noted that hybrid coatings have greater thickness than inorganic coatings, so their effectiveness against corrosion is usually greater. The greater thickness of these coatings is due to the presence of residual internal porosity, generated by the non-hydrolyzable organic groups of the structural network of the coating [56]. These pores are closed and are not detrimental to the anticorrosive behavior of the coating, although they do significantly reduce their mechanical behavior.

Currently, it is sought that the sol-gel coatings on aluminum alloys, in addition to having a good corrosion behavior, also have a good mechanical behavior. The mechanical properties of the coatings made by the sol-gel route are not easy to determine; the modulus of elasticity, the hardness, the adhesion of the coatings to the metallic substrates, and the tribological properties (wear) are the main properties that have been evaluated in this type of coatings.

4.1. Mechanical properties: determination of modulus of elasticity, hardness, and fracture toughness of coatings

Parameters such as modulus of elasticity (E), hardness (H), or fracture toughness must be known to anticipate the in-service performance of such coatings. The most used technique for the determination of the modulus of elasticity of materials is the tensile test, which is not applicable to characterize a coating. The coatings generated by sol-gel have thickness in the order of microns, so that the usual techniques of characterization of the hardness of the materials, hardness or microhardness, apply too much load to the coatings, obtaining the mechanical response of the substrate also. The main obstacle that exists when knowing the mechanical properties of the coatings by hardness tests is to avoid the influence of the substrate on the results of the test, which leads to perform tests at micro- or nanometric scale, depending on the thickness of the covering.

The microhardness test instruments (micro-durometers) do not allow to apply forces small enough to provide penetrations of the order of 10% of the thickness of the coating, necessary to avoid the influence of the substrate in the measurements made, an essential factor when the coating has small thickness. In addition, the durometers base the determination of the hardness in the measurement of the size of the residual footprint left by the penetrator on the surface tested, but at such a low load, to achieve low penetration, this residual trace cannot be determined with sufficient accuracy as to provide acceptable hardness values. As an example, the uncertainty associated with the determination, by conventional optical methods, of a diagonal measuring 5 μm corresponding to the residual footprint made with Vickers indenter is of the order of 20%. This uncertainty increases as the size of the diagonal decreases, being able to reach 100% for a size of 1 μm.

This leads to the need of developing new mechanical characterization techniques for thin coatings. Among them the most used, and that allows the determination of both *E* and *H*, is nanoindentation. The nanoindentation technique overcomes the limitation of the measurement of the size of the footprint basing the determination of the hardness (*H*) and the modulus of elasticity (*E*) of the material in continuous measurement of the depth of penetration and the known geometry of the indentator.

In this technique, the applied load displacement curve inside the material is recorded continuously. In the initial part of the load cycle, at low applied load, the tested material elastically deforms, to become plastically deformed at higher loads. If the plasticization of the material has taken place during the loading process, the load-displacement data of the discharge branch are different from those of the load branch. In this way, a trace is generated on the surface of the material tested, because the plastic deformation generated has not been recovered, only the elastic deformation. The Berkovich indentator is the one commonly used in nanoindentation tests, because it has a three-sided pyramid geometry in which it is easier to achieve a point vertex than with a four-sided pyramid (Vickers), allowing better control of the process of indentation.

The nanoindentation tests in this type of coatings are generally carried out by means of a nanoindentation module coupled to an AFM equipment, avoiding the mechanical response of the substrate. This allows having the resolution of the AFM in the horizontal and vertical displacement and therefore carrying out the tests in the selected areas with high precision.

The mechanical properties of the coating, as well as its resistance to corrosion, are also modified by the densification temperature used, since it conditions the microstructure of the obtained coating, being able to go from an amorphous state to a crystalline state. Olonfinjama and collaborators [57] proved the improvement of the mechanical properties obtained in mononane and multilayer titania coatings with crystalline microstructure (densification at 500°C) deposited on metal substrates, with respect to obtaining amorphous microstructure (densification at room temperature). The results obtained by nanoindentation at very low load show that the obtaining of crystalline coatings implies a 25% increase in the hardness of the coating (1.5 GPa) and an increase of approximately 40% of the modulus of elasticity (85 GPa) with respect to the coating values in the amorphous state. This shows that by means of the heat treatment at high temperature, the coating has gone from an amorphous initial state to a crystalline structure, beneficial for the mechanical performance of the coating. The influence of the thickness of the coatings in the mechanical properties of these is null.

By means of thermal treatments at a high-temperature furnace, the densification of the coatings is achieved, although there are other ways to achieve this densification. The influence of the densification technique on the mechanical properties of the coating is evident in the research carried out by Jämting et al. [58], which densify titania sol-gel coatings by bombardment with hydrogen ions and by heating in an oven. The nanoindentation technique demonstrates that by bombardment the highest densifications are obtained in the coating in contact with the substrate, while densifying in the furnace the greater densifications of the coating is achieved in the surface area. The time used in the densification also modifies the mechanical properties of the coating, as Lucca et al. [59] confirm in zirconia coatings made by sol-gel and coatings by immersion on metal substrates.

The nanoindentation technique was also used to calculate the fracture toughness of coatings [60], since the load-displacement curves obtained from the tests make it possible to determine the load at which the coating cracks.

Mammeri et al. [61] investigated the mechanical properties of hybrid silica coatings by nanoindentation, demonstrating that the test discharge curve does not reflect only the elastic properties of the coating but shows the creep induced by the response of the polymeric zones of the coating. Therefore, the time in which the discharge is performed must be designed to avoid this temporary response of the coating as much as possible, making a series of corrections [62] for the calculation of the modulus of elasticity and the hardness of the material.

A novel way of obtaining and densifying sol-gel coatings are by using the laser technique [63] with which coatings with high values of E and H are obtained, possibly due to the refinement by laser densification of the structure of the obtained coatings.

4.2. Adherence: determination of the adhesion of coatings

The usual techniques for determining the adhesion of coatings such as three-point bending techniques or the technique of peeling with adhesive tapes lose effectiveness when evaluating the adhesion of fine ceramic coatings. This is normally because the failure of these coatings is due to cracking, since they are fragile coatings.

Techniques such as nanoindentation or nanoray are being used to determine the adhesion of this type of coatings, including coatings obtained by sol-gel. In nanoindentation tests, cracking at the interface is detected in the load-displacement curve since a change in slope occurs during the loading process. By means of the nanoray tests, in which the normal load applied to the material increases while the indenter moves over the surface of a series of microns, the loads can be detected at which the separation between coating and substrate occurs, either by acoustic methods, by sudden increase in the coefficient of friction, or by the subsequent observation of the scratching track.

The surface roughness of the substrate and the densification temperature of the coating are factors that influence the adherence of the coatings. Xie and Hawthorne [64] show that the adhesion of the sol-gel coatings increases with increasing surface roughness of the substrates and the densification temperature. When the generated sol-gel coating is hybrid, increasing the proportion of the non-hydrolyzable alkoxide increases the adhesion of the coating to the substrate [65].

Another way to determine the adhesion between coating and substrate is the use of traction tests on pieces joined to a simple overlap using an epoxy base adhesive.

4.3. Tribological properties: weathering

So that the coating can be used in anti-wear applications, it must have thicknesses between 0, 5, and 10 μm, with which multilayer systems are used when the sol-gel route is chosen to manufacture the said coatings. Normally, high temperatures are used for the densification of the coating, so it can meet the anti-wear requirements [65]. The temperature must be selected considering that the mechanical properties of the substrate do not decrease. This is especially important when working with substrates of aluminum alloys, since the temperatures at which this change in properties occurs are much lower than in the case of titanium alloys or carbon steels. The densification temperatures influence the final structure of the coating. Thus, high temperatures tend to form crystalline coatings, while low temperatures tend to form amorphous coatings.

Sol-gel coatings for anti-wear applications are usually fundamentally inorganic, with the most common being those of alumina, zirconia, or silica. The use of hybrid coatings is less widespread, due to the mechanical limitations that often appear in these coatings because of their high percentage of porosity. However, the use of modified inorganic coatings is extended, either by the addition of lubricating particles that reduce the coefficients of friction or by the addition of organic modifiers to the starting sol that generate a decrease in the roughness of the coating.

Taktak and Baspinar [66] demonstrated an augment of the wear resistance by increasing the crystalline and decreasing of the coefficient of friction. These effects were explained based on two concepts: First, the presence of crystalline phase in an amorphous matrix prevents the propagation of cracks originated during the wear process, due to the presence of crystalline grain boundaries [67]. The presence of crystalline phase in an amorphous matrix increases the strength and the fracture tenacity of the material, due to the compression stresses that the said phase generates [68].

The doping of hard coatings is another of the widely ways used to improve their mechanical or tribological properties [69].

A typical way to evaluate the wear behavior of the coating is through *pin-on-disc* tests without lubrication and at room temperature. Once the tests have been carried out, the wear tracks are observed by means of SEM in order to correlate the values obtained after the tests with the morphology of the wear tracks.

5. Conclusions

The coating obtained using sol-gel processing has shown good performance as corrosion barrier in the protection of metal substrate. The versatility along with the "green" methodology makes this process an excellent alternative to replace the conventional coating.

The hybrid polymer improves the mechanical properties and allows a better control in the preparation of coat. Moreover, the process to obtain the polymer allows the incorporation of organic and inorganic compounds. Thus, considering these points, the effort of the scientific community is obtaining a "smart coating," which present multiple properties.

Acknowledgements

The authors thank FONDECYT (Grant 11170419), PIA-CONICYT (Grant ACT-1412), DICYT-USACH (051742PC_DAS), and AFOSR (Grant FA 9550-16-1-0063) for financial support.

Author details

Evelyn Gonzalez[1], Nelson Vejar[2], Roberto Solis[2], Lisa Muñoz[1], Maria Victoria Encinas[3] and Maritza Paez[1]*

*Address all correspondence to: maritza.paez@usach.cl

1 Department of Materials Science, Faculty of Chemistry and Biology, University of Santiago of Chile, Estación Central, Chile

2 Aerospace Sciences Research and Development Centre (CIDCA), Chilean Air Force, San Bernardo, Chile

3 Department of Environment Science, Faculty of Chemistry and Biology, University of Santiago of Chile, Estación Central, Chile

References

[1] Garcia-Heras M, Jimenez-Morales A, Casal B, Galvan JC, Radzki S, Villegas MA. Preparation and electrochemical study of cerium-silica sol-gel thin films. Journal of Alloys and Compounds. 2004;**380**:219-224. DOI: 10.1016/j.jallcom.2004.03.047

[2] Shao M, Fu Y, Hu R, Lin C. A study on pitting corrosion of aluminum alloy 2024-T3 by scanning microreference electrode technique. Materials Science and Engineering A. 2003;**344**:323-327. DOI: 10.1016/S0921-5093(02)00445-8

[3] Capelossi VR, Poelman M, Recloux I, Hernandez RPB, de Melo HG, Olivier MG. Corrosion protection of clad 2024 aluminum alloy anodized in tartaric-sulfuric acid bath

and protected with hybrid sol-gel coating. Electrochimica Acta. 2014;**124**:69-79. DOI: 10.1016/j.electacta.2013.09.004

[4] Xia L, McCreery R. Chemistry of a chromate conversion coating on aluminum alloy AA2024-T3 probed by vibrational spectroscopy. Journal of the Electrochemical Society. 1998;**145**:3083-3089. DOI: 10.1149/1.1838768

[5] Zhao J. Corrosion protection of untreated AA-2024-T3 in chloride solution by a chromate conversion coating monitored with Raman spectroscopy. Journal of the Electrochemical Society. 1998;**145**:2258. DOI: 10.1149/1.1838630

[6] Metroke TL, Parkhill RL, Knobbe ET. Passivation of metal alloys using sol-gel-derived materials—A review. Progress in Organic Coatings. 2001;**41**:233-238. DOI: 10.1016/S0300-9440(01)00134-5

[7] Brinker CJ, Hurd AJ, Schunk PR, Frye GC, Ashley CS. Review of sol-gel thin film formation. Journal of Non-Crystalline Solids. 1992;**147-148**:424-436. DOI: 10.1016/S0022-3093(05)80653-2

[8] Wright J, Sommerdijk NAJM. Sol-Gel Materials: Chemistry and Applications. Amsterdam: CRC Press; 2001

[9] Donley M, Knobbe E. Private communication. n.d

[10] Brinker CJ, Hurd AJ, Frye GC, Ward KJ, Ashley CS. Sol-gel thin film formation. Journal of Non-Crystalline Solids. 1990;**121**:294-302. DOI: 10.1016/0022-3093(90)90147-E

[11] Niederberger M, Pinna N. Metal oxide nanoparticles in organic solvents: Synthesis, formation, assembly and application. History. London: Springer-Verlag; 2009:7-18. DOI: 10.1007/978-1-84882-671-7

[12] Brinker CJ, Scherer GW. Sol-Gel Science, The Physics and Chemistry of Sol-Gel Processing. 1st ed. Boston: Academic Press LTDA; 1991. DOI: 10.1002/adma.19910031025

[13] Brinker CJ. Hydrolysis and condensation of silicates: Effects on structure. Journal of Non-Crystalline Solids. 1988;**100**:31-50

[14] Abdolah Zadeh M, van der Zwaag S, Garcia SJ. Routes to extrinsic and intrinsic self-healing corrosion protective sol-gel coatings: a review. Self-Healing Mater. 2013;**1**:1-18. DOI: 10.2478/shm-2013-0001

[15] Zheludkevich ML, Miranda I, Ferreira MGS. Sol-gel coatings for corrosion protection of metals. Journal of Materials Chemistry. 2005;**15**:5099-5111. DOI: 10.1039/b419153f

[16] Livage J, Henry M, Sanchez C. Sol-gel chemistry of transition metal oxides. Progress in Solid State Chemistry. 1988;**18**:259-341. DOI: 10.1016/0079-6786(88)90005-2

[17] Hench LL, West JK. The sol-gel process. Chemical Reviews. 1990;**90**:33-72. DOI: 10.1021/cr00099a003

[18] Parkhill RL, Knobbe ET, Donley MS. Application and evaluation of environmentally compliant spray-coated ormosil films as corrosion resistant treatments for aluminum 2024-T3. Progress in Organic Coatings. 2001;**41**:261-265. DOI: 10.1016/S0300-9440(01)00138-2

[19] Veeraraghavan B, Haran B, Slavkov D, Prabhu S, Popov B, Heimann B. Development of a novel electrochemical method to deposit high corrosion resistant silicate layers on metal substrates. Electrochemical and Solid-State Letters. 2003;**6**:B4. DOI: 10.1149/1.1537092

[20] Ogawa M. A simple sol-gel route for the preparation of silica-surfactant mesostructured materials. Chemical Communications. 1996;**0**:1149-1150

[21] Ikawa M, Yamada T, Matsui H, Minemawari H, Tsutsumi J, Horii Y, et al. Simple push coating of polymer thin-film transistors. Nature Communications. 2012;**3**:1-8. DOI: 10.1038/ncomms2190

[22] Tyona MD. A theoretical study on spin coating technique. Advanced Material Research. 2013;**2**:195-208. DOI: 10.12989/amr.2013.2.4.195

[23] Faustini M, Louis B, Albouy PA, Kuemmel M, Grosso D. Preparation of sol-gel films by dip-coating in extreme conditions. Journal of Physical Chemistry C. 2010;**114**:7637-7645. DOI: 10.1021/jp9114755

[24] Sheffer M, Groysman A, Mandler D. Electrodeposition of sol-gel films on Al for corrosion protection. Corrosion Science. 2003;**45**:2893-2904. DOI: 10.1016/S0010-938X(03)00106-9

[25] Figueira RB, Silva CJR, Pereira EV. Organic-inorganic hybrid sol-gel coatings for metal corrosion protection: A review of recent progress. 2015;**12**:1-35. DOI: 10.1007/s11998-014-9595-6

[26] Wang D, Bierwagen GP. Sol-gel coatings on metals for corrosion protection. Progress in Organic Coatings. 2009;**64**:327-338. DOI: 10.1016/j.porgcoat.2008.08.010

[27] van Ooij WJ, Zhu D, Stacy M, Seth A, Mugada T, Gandhi J, et al. Corrosion protection properties of organofunctional silanes—An overview. Tsinghua Science and Technology. 2005;**10**:639-664. DOI: 10.1016/S1007-0214(05)70134-6

[28] Li Q, Yan Y, Yu M, Song B, Shi S, Gong Y. Synthesis of polymeric fluorinated sol-gel precursor for fabrication of superhydrophobic coating. Applied Surface Science. 2016;**367**:101-108. DOI: 10.1016/j.apsusc.2016.01.155

[29] Vivar Mora L, Taylor A, Paul S, Dawson R, Wang C, Taleb W, et al. Impact of silica nanoparticles on the morphology and mechanical properties of sol-gel derived coatings. Surface and Coatings Technology. 2018;**342**:48-56. DOI: 10.1016/j.surfcoat.2018.02.080

[30] Wu G, An J, Sun D, Tang X, Xiang Y, Yang J. Robust microcapsules with polyurea/silica hybrid shell for one-part self-healing anticorrosion coatings. Journal of Materials Chemistry A. 2014;**2**:11614-11620. DOI: 10.1039/C4TA01312C

[31] Zheludkevich ML, Serra R, Montemor MF, Miranda Salvado IM, Ferreira MGS. Corrosion protective properties of nanostructured sol-gel hybrid coatings to AA2024-T3. Surface and Coatings Technology. 2006;**200**:3084-3094. DOI: 10.1016/j.surfcoat.2004.09.007

[32] Zheludkevich ML, Serra R, Montemor MF, Yasakau KA, Salvado IMM, Ferreira MGS. Nanostructured sol-gel coatings doped with cerium nitrate as pre-treatments for AA2024-T3 Corrosion protection performance. Electrochimica Acta. 2005;**51**:208-217. DOI: 10.1016/j.electacta.2005.04.021

[33] Khramov AN, Voevodin NN, Balbyshev VN, Mantz RA. Sol-gel-derived corrosion-protective coatings with controllable release of incorporated organic corrosion inhibitors. Thin Solid Films. 2005;**483**:191-196. DOI: 10.1016/j.tsf.2004.12.021

[34] Yasakau KA, Zheludkevich ML, Lamaka SV, Ferreira MGS. Mechanism of corrosion inhibition of AA2024 by rare-earth compounds. The Journal of Physical Chemistry. B. 2006;**110**:5515-5528. DOI: 10.1021/jp0560664

[35] De Nicolò A, Paussa L, Gobessi A, Lanzutti A, Cepek C, Andreatta F, et al. Cerium conversion coating and sol-gel multilayer system for corrosion protection of AA6060. Surface and Coatings Technology. 2016;**287**:33-43. DOI: 10.1016/j.surfcoat.2015.12.059

[36] Vreugdenhil AJ, Woods ME. Triggered release of molecular additives from epoxy-amine sol-gel coatings. Progress in Organic Coatings. 2005;**53**:119-125. DOI: 10.1016/j.porgcoat.2005.02.004

[37] Roussi E, Tsetsekou A, Skarmoutsou A, Charitidis CA, Karantonis A. Anticorrosion and nanomechanical performance of hybrid organo-silicate coatings integrating corrosion inhibitors. Surface and Coatings Technology. 2013;**232**:131-141. DOI: 10.1016/j.surfcoat.2013.04.063

[38] Maia F, Yasakau KA, Carneiro J, Kallip S, Tedim J, Henriques T, et al. Corrosion protection of AA2024 by sol-gel coatings modified with MBT-loaded polyurea microcapsules. Chemical Engineering Journal. 2016;**283**:1108-1117. DOI: 10.1016/j.cej.2015.07.087

[39] Shchukina E, Shchukin D, Grigoriev D. Effect of inhibitor-loaded halloysites and mesoporous silica nanocontainers on corrosion protection of powder coatings. Progress in Organic Coatings. 2017;**102**:60-65. DOI: 10.1016/j.porgcoat.2016.04.031

[40] Kirk-Othmer Encyclopedia of Chemical Technology. Wiley and Sons. 2000. pp 1-20. Available from: http://pubs.acs.org/doi/full/10.1021/acsreagents.2008

[41] Vejar ND, Azocar MI, Tamayo LA, Gonzalez E, Gulppi M, Zhou X, et al. Paez MA, Centre P. Antibiofouling properties of sol-gel type polymers for aluminium alloys: Biocorrosion protection against *Pseudomonas Aeruginosa*. International Journal of Electrochemical Science. 2013;**8**:12062-12077

[42] Fir M, Orel B, Vuk AS, Vilcnik A, Jese R, Francetic V. Corrosion studies and interfacial bonding of urea/poly(dimethylsiloxane) sol/gel hydrophobic coatings on AA 2024 aluminum alloy. Langmuir. 2007;**23**:5505-5514. DOI: 10.1021/la062976g

[43] Varma PCR, Colreavy J, Cassidy J, Oubaha M, McDonagh C, Duffy B. Corrosion protection of AA 2024-T3 aluminium alloys using 3, 4-diaminobenzoic acid chelated zirconium-silane hybrid sol-gels. Thin Solid Films. 2010;**518**:5753-5761. DOI: 10.1016/j.tsf.2010.05.088

[44] Arnott DR, Ryan NE, Hinton BRW, Sexton BA, Hughes AE. Auger and XPS studies of cerium corrosion inhibition on 7075 aluminum alloy. Applied Surface Science. 1985;**22-23**:236-251. DOI: 10.1016/0378-5963(85)90056-X

[45] Van der Heide P. Introduction, X-Ray Photoelectron Spectroscopy: An Introduction to Principles and Practices. Wiley & Sons, Wienhein; 2012. pp. 1-12. https://doi.org/10.1002/anie.201205395

[46] Kasten LS, Grant JT, Grebasch N, Voevodin N, Arnold FE, Donley MS. An XPS study of cerium dopants in sol-gel coatings for aluminum 2024-T3. Surface and Coatings Technology. 2001;**140**:11-15. DOI: 10.1016/S0257-8972(01)01004-0

[47] Chen M, Wang X, Yu YH, Pei ZL, Bai XD, Sun C, et al. X-ray photoelectron spectroscopy and auger electron spectroscopy studies of Al-doped ZnO films. Applied Surface Science. 2000;**158**:134-140. DOI: 10.1016/S0169-4332(99)00601-7

[48] Mehta R. Interactions, imaging and spectra in SEM. In: Scanning Electron Microscopy. Intech; 2012:17-30

[49] Yasakau KA, Kallip S, Zheludkevich ML, Ferreira MGS. Active corrosion protection of AA2024 by sol-gel coatings with cerium molybdate nanowires. Electrochimica Acta. 2013;**112**:236-246. DOI: 10.1016/j.electacta.2013.08.126

[50] García-Heras M, González-Mena E, Galvín JC, Villegas MA. Modificación superficial de aleaciones de base aluminio (anodizadas y no anodizadas) mediante recubrimientos de sílice. Revista de Metalurgia. 2004;**40**:127-131

[51] Hamdy AS, Butt DP. Environmentally compliant silica conversion coatings prepared by sol-gel method for aluminum alloys. Surface and Coatings Technology. 2006;**201**: 401-407. DOI: 10.1016/j.surfcoat.2005.11.142

[52] Voevodin NN, Balbyshev VN, Donley MS. Investigation of corrosion protection performance of sol-gel coatings on AA2024-T3. Progress in Organic Coatings. 2005;**52**:28-33. DOI: 10.1016/j.porgcoat.2004.05.006

[53] Liu Y, Sun D, You H, Chung JS. Corrosion resistance properties of organic-inorganic hybrid coatings on 2024 aluminum alloy. Applied Surface Science. 2005;**246**:82-89. DOI: 10.1016/j.apsusc.2004.10.040

[54] Liu L, Hu JM, Zhang JQ, Cao CN. Improving the formation and protective properties of silane films by the combined use of electrodeposition and nanoparticles incorporation. Electrochimica Acta. 2006;**52**:538-545. DOI: 10.1016/j.electacta.2006.05.034

[55] Conde A, Durán A, De Damborenea JJ. Polymeric sol-gel coatings as protective layers of aluminium alloys. Progress in Organic Coatings. 2003;**46**:288-296. DOI: 10.1016/S0300-9440(03)00014-6

[56] Pepe A, Aparicio M, Ceré S, Durán A. Synthesis of hybrid silica sol-gel coatings containing Zn particles on carbon steel and Al/Zn coated carbon steel. Materials Letters. 2005;**59**:3937-3940. DOI: 10.1016/j.matlet.2005.07.037

[57] Olofinjana AO, Bell JM, Jamting AK. Evaluation of the mechanical properties of sol-gel-deposited titania films using ultra-micro-indentation method. Wear. 2000;**241**: 174-179. DOI: 10.1016/S0043-1648(00)00372-0

[58] Jämting ÅK, Bell JM, Swain MV, Wielunski LS, Clissold R. Measurement of the micro mechanical properties of sol-gel TiO_2 films. Thin Solid Films. 1998;**332**:189-194. DOI: 10.1016/S0040-6090(98)01102-X

[59] Lucca DA, Klopfstein MJ, Ghisleni R, Gude A, Mehne A, Datchary W. Investigation of sol-gel derived ZrOz thin films by nanoindentation. CIRP Annals—Manufacturing Technology. 2004;**53**:475-478

[60] Malzbender J, De With G. Elastic modulus, hardness and fracture toughness of SiO_2-filled methyltrimethoxysilane coatings on glass substrates. Journal of Non-Crystalline Solids. 2000;**265**:51-60. DOI: 10.1016/S0022-3093(99)00889-3

[61] Mammeri F, Le Bourhis E, Rozes L, Sanchez C. Elaboration and mechanical characterization of nanocomposites thin films. Part I: Determination of the mechanical properties of thin films prepared by in situ polymerisation of tetraethoxysilane in poly(methyl methacrylate). Journal of the European Ceramic Society. 2006;**26**:259-266. DOI: 10.1016/j.jeurceramsoc.2004.11.013

[62] Tang B, Ngan AHW. Accurate measurement of tip-sample contact size during nanoindentation of viscoelastic materials. Journal of Materials Research. 2003;**18**:1141-1148. Available form: http://hdl.handle.net/10722/156673%0AJournal

[63] Ezz T, Crouse P, Li L, Liu Z. Combined laser/sol-gel synthesis of Si/O/C coatings on mild steel. Surface and Coatings Technology. 2006;**200**:6395-6399. DOI: 10.1016/j.surfcoat.2005.11.080

[64] Xie Y, Hawthorne HM. Measuring the adhesion of sol-gel derived coatings to a ductile substrate by an indentation-based method. Surface and Coatings Technology. 2003;**172**:42-50. DOI: 10.1016/S0257-8972(03)00319-0

[65] Hübert T, Svoboda S, Oertel B. Wear resistant alumina coatings produced by a sol-gel process. Surface and Coatings Technology. 2006;**201**:487-491. DOI: 10.1016/j.surfcoat.2005.11.014

[66] Taktak S, Baspinar MS. Wear and friction behaviour of alumina/mullite composite by sol-gel infiltration technique. Materials and Design. 2005;**26**:459-464. DOI: 10.1016/j.matdes.2004.07.012

[67] Medvedovski E. Wear-resistant engineering ceramics. Wear. 2001;**249**:821-828. DOI: 10.1016/S0043-1648(01)00820-1

[68] Marple BR, Green DJ. Mullite/Alumina particulate composites by infiltration. Processing: II, Infiltration and Characterization. Journal of the American Ceramic Society. 1990;**73**:3611-3616

[69] Donnet C. Recent progress on the tribology of doped diamond-like and carbon alloy coatings: a review. Surface and Coatings Technology. 1998;**100-101**:180-186. DOI: 10.1016/S0257-8972(97)00611-7